DOODLEBUGS & DOWSERS

DOODLEBUGS & DOWSERS

A HISTORY OF UNUSUAL WAYS TO SEARCH FOR OIL

DAN PLAZAK

TECH TECH UNIVERSITY PRESS

This book is typeset in EB Garamond. The paper used in this book meets the minimum requirements of ANSI/NISO Z39.48-1992 (R1997). ⊗

Designed by Hannah Gaskamp
Cover design by Hannah Gaskamp

Library of Congress Cataloging-in-Publication Data

Names: Plazak, Dan, 1951– author. Title: Doodlebugs and Dowsers: A History of Unusual Ways to Search for Oil / Dan Plazak. Description: Lubbock, Texas: Texas Tech University Press, [2023] | Includes bibliographical references and index. Identifiers: LCCN 2023001601 (print) | LCCN 2023001602 (ebook) | ISBN 978-1-68283-177-9 (paperback) | ISBN 978-1-68283-178-6 (ebook) Subjects: LCSH: Petroleum—Prospecting—United States. | Dowsing—United States. | Swindlers and swindling—United States. Classification: LCC TN271.P4 P53 2023 (print) | LCC TN271.P4 (ebook) | DDC 622/.182820973—dc23/eng/20230301 LC record available at https://lccn.loc.gov/2023001601 LC ebook record available at https://lccn.loc.gov/2023001602

Texas Tech University Press
Box 41037
Lubbock, Texas 79409-1037 USA
800.832.4042
ttup@ttu.edu
www.ttupress.org

CONTENTS

ILLUSTRATIONS   vii

CHAPTER 1: AN OIL INDUSTRY WITHOUT
GEOLOGISTS   3

CHAPTER 2: ALL I HAVE TO DO IS DREAM   7

CHAPTER 3: PETROLEUM SPIRITS   11

CHAPTER 4: THE PSYCHICS   21

CHAPTER 5: INTRODUCTION TO DOWSING   63

CHAPTER 6: EARLY OIL DOWSING   67

CHAPTER 7: OIL DOWSING IN THE AGE OF
GEOLOGY   81

CHAPTER 8: A WORLD OF OIL DOWSERS   109

CHAPTER 9: INTRODUCTION TO
DOODLEBUGS   123

CHAPTER 10: WHEN DOODLEBUGS RULED
THE EARTH   135

CHAPTER 11: DOODLEBUGS AROUND
THE WORLD   181

CHAPTER 12: TWILIGHT OF THE
DOODLEBUGS   199

CHAPTER 13: LOOKING BACK,
LOOKING AHEAD   237

CONTENTS

NOTES   245

REFERENCES   287

INDEX   289

ILLUSTRATIONS

4   Fig. 1. Radionics device designed by Wesley C. Miller

22   Fig. 2. Stella Teller being kidnapped to return her to the insane asylum

26   Fig. 3. Graves of Annie Webb and husband, Rev. Clark West

29   Fig. 4. Areas of Cayce Petroleum activity

44   Fig. 5. Daisy Bradford #3 oil well

68   Fig. 6. An early oil dowser, also known as an oil witch

69   Fig. 7. John Booher's well locations

71   Fig. 8. John Booher searching for gas

74   Fig. 9. Bobber of oil dowser Andrew Thompson

80   Fig. 10. Discovery well for the Boulder oil field

82   Fig. 11. Anticline sketch

86   Fig. 12. Newspaper cartoon of Henry Zachary

105   Fig. 13. Jacob Long, known as the "oil wizard"

126   Fig. 14. Map of localities in southeast Texas

136   Fig. 15. A satirical view of doodlebugs

143   Fig. 16. A cartoonist's view of the Mexia boom

150   Fig. 17. Drilled locations in Oregon selected by Reverend Olson

168   Fig. 18. Drilled locations by dowsing and doodlebugging

175   Fig. 19. L. V. J. Kimball

182   Fig. 20. J. Lyle Telford

191   Fig. 21. The Schermuly Polarizer

195   Fig. 22. Advertisement for the Mansfield Automatic Water & Oil Finder

196   Fig. 23. The face of the Mansfield detector

201   Fig. 24. Oil-dowsing equipment used by James M. Young

203   Fig. 25. Wildcatter Assaph "Ace" Gutowsky

238   Fig. 26. Number of known oil dowsers in the United States

239   Fig. 27. Number of known oil doodlebuggers in the United States
240   Fig. 28. Number of new doodlebuggers by decade

DOODLEBUGS

&

DOWSERS

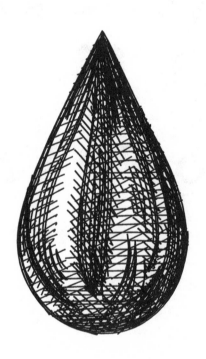

AN OIL INDUSTRY WITHOUT GEOLOGISTS

THE BLACK BOX IN CALGARY

THE MAN WITH THE "BLACK BOX" USED TO BE A STAPLE OF OIL patch stories. The box is an oil-finding gizmo about the size of a cake box, with dials and gauges on the top. The man explains that it is all very scientific—something that you can't quite follow about gravity waves, or sunspots, or cosmic rays, although the details of its construction are a carefully guarded secret. He has tested it over buckets of used motor oil, and next to pumping oil wells, and it is infallible—of that he is certain. He drives out to the drill rig and sits in his car fiddling with the dials, watching the gauges, then steps out to announce that the well will yield oil or will be a dry hole. And maybe he's right.

The fellow in Calgary, Canada's oil capital, was an old man, but no wild-eyed inventor. Ned Gilbert was a retired petroleum land man, well known in the Calgary oil patch, and the oil-finding machine was the oddest memento of his long and successful career. He brought out his black box and laid it on a table in his high-rise condo.

Ned lent me a screwdriver and let me open the black metal box to discover its secrets. He said that it came from California in the 1950s, and years ago there were a number of machines of this model around Calgary, used to search for oil in Western Canada. It looked like professionally assembled circuitry from

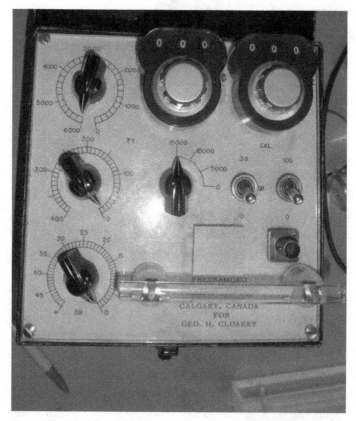

Fig. 1. The classic black box, a Radionics device designed and used by sound engineer Wesley C. Miller and later used by George S. Hume, ex-head of the Geological Survey of Canada. (Photo by the author.)

that era. Dials and knobs decorated the top of the metal box. Inside were typical electrical components, such as variable resistors and capacitors. There was no place for a battery, however, and no outlet to a power source. By all we know of electricity, this neatly wired circuitry was useless.

Ned could demonstrate how to operate the machine, even though he couldn't explain it. He wasn't selling anything, just letting me examine the gizmo as a favor. He named a top petroleum geologist who had successfully used it to find oil.

The black box was a doodlebug.

Since the early 1900s, "doodlebug" has been oil-industry slang for an unscientific instrument that, according to its maker, can find oil deposits with magic ease and unerring certainty. As with many slang terms, doodlebug is often used carelessly, but it is used here in its original meaning: a pseudo-geophysical instrument, machines pretending to a scientific basis.

BEFORE GEOLOGISTS

Today geologists are synonymous with oil exploration, but it was not always so. There was a time before geologists. The American oil industry started in 1859 at Titusville, Pennsylvania, but for decades, practical oil men sneered that geology never found a barrel of oil.

It was a different world then. Touring the peaceful rolling hills of northwest Pennsylvania, it's difficult to imagine the grit and fury of the world's first oil boom 160 years ago. Gone is the pervasive smell of crude oil. The hillsides were then barren of all but stumps after the trees had been chopped down and reassembled into forests of wooden oil derricks. Today the green forest is back. Oil Creek, once crowded with derricks and black with spilled petroleum, is again a sparkling country stream. Oil City, which a visitor in 1865 described as "that perfection of filth and disorder," is today a neat and pleasant town of beautiful period buildings.[1]

It was not only the physical environment that was different but also the realm of knowledge. Oil drilling was a risky business, and science at first had little to offer. Dry holes were common, and many wells made only a dribble of oil—not enough to pay back expenses. Oil drillers were desperate to reduce the risk, and many turned to hunches, dreams, dowsing, and apparitions from the spirit world.

Drillers resorted to rank superstitions. Drilling locations were marked by driving a stake of green wood into the ground: to use dry wood would produce a dry hole. A cross-eyed ox hauling the drill rig to the new location would likewise condemn the hole to failure. Drillers would not ride blaze-faced horses because doing so caused problems with the drilling. Holes were never started on Fridays or on the thirteenth of the month. A dead skunk thrown down the hole would guarantee an oil well.[2] Even those who were not superstitious might act in accordance with superstitions because, well, it wouldn't hurt to be safe.

Oil drilling spread to other states and continued to rely on rules of thumb. Cemeteries were considered good places to find oil. Some people looked for certain vegetation. In Oklahoma, Harry Sinclair favored blackjack oak.[3] There was the rumored oil flower, the identity of which was a secret, but it bloomed only above oil pools.

The surface usually gives little clue to the oil hidden below. The cattle pasture barren of oil looks much the same as the neighbor's land that covers a fabulous oil field. A former tropical coral reef underlies today's snow-covered pine forest. An ancient beach thousands of feet underground and hundreds of miles from the coast marks the edge of a long-ago continent. The gravel riverbed of a hundred million years ago is buried deep but still winds "down to a sunless sea." Drill a well into an ancient riverbed and you may find it saturated with oil—but drill a few hundred feet away and you could miss it completely and have no inkling of what was lost.

Geologists were scarce in the oil business for its first half-century. Today, geologists choose almost every location drilled. But most scientific methods of exploration are time-consuming, expensive, and too often still risky. All the scientific advancements have not tamed that awful thing called chance. Exploring for oil and gas is terribly inefficient, and none are more sadly and acutely aware of this than geologists. There must be a better way.

There are exploration methods that boast astounding success rates and are usually much cheaper and faster than the scientific approaches. What methods are these? They are precisely those that geology replaced. Geologists and geophysicists often assume that science long ago swept the old practices into the dustbin of superstition—but they are wrong.

Oil dowsers, although reduced in number, still search with rod and pendulum. Psychics still offer to find oil. Oil exploration will always be a human endeavor, and while the old methods are short on science, they are rich in humanity. These unscientific methods tempt us on a deeply human level. That dream seemed so real, that dowser so certain; what if he is on to something? Can this psychic really find oil? Did the inventor of that crazy gizmo stumble on a better way to find oil?

ALL I HAVE TO DO IS DREAM

SINCE ANCIENT TIMES, PEOPLE HAVE SEEN DREAMS AS WINDOWS to the future. Dreams reflect subconscious and conscious desires, so it is natural that people dream of finding wealth in an oil well. Dreams may raise false hopes, but sometimes dreams come true.

THE COQUETTE

The first documented and most famous dream of a potential oil well was the Coquette. According to George M. Kepler, in 1864 he took a young lady he was wooing to a ceremonial dance by a group of Indians in Hydetown, Pennsylvania. Afterward, he asked the woman to marry him, but she turned him down. That night, Kepler dreamt that an Indian was preparing to shoot him with an arrow when the young lady who had rejected him handed him a rifle. He fired the rifle at the Indian, and the Indian disappeared, his place taken by a stream of crude oil gushing from the ground.

The dream seemed like a jumble of everything on Kepler's mind: Indians, frustrated romance, and petroleum. Petroleum was on his mind because the next day he was going to Petroleum Center, Pennsylvania, where his cousin Aaron C. Kepler was the superintendent of the Hyde and Egbert Farm, a prolific oil property. According to George Kepler, when he arrived at the farm, he recognized the place in his dream where the Indian had turned into a fountain of oil. He and three others leased one acre surrounding the location he saw in his dream and agreed to pay the landowners an unheard-of 75 percent royalty. The partners built an oil derrick on the dream spot and named their well the Coquette, after the woman of Kepler's dream. The well finished drilling in the spring of 1864 and started pumping oil. But when the owners pulled out the pump to make a repair, the well gushed 1,200 barrels per day. George Kepler sold his share after three months, making a profit of $80,000.

According to one version of the story, George returned home, having made his fortune on the Coquette well, and proposed once more to the lady of his dreams; this time she accepted, and they married. The success of the Coquette well, and its romantic story, made a sensation. As if they weren't making enough money from the oil, the owners charged tourists ten cents to see the well.[1]

The Coquette is a great story and is recounted in numerous books. But a well drilled anywhere on the Hyde and Egbert Farm could not have missed finding oil. Many years later, A. C. Kepler recalled a different version of events: "After tossing a hat in the air, and noting the spot where it fell to indicate where we should start to drill, we located our well, and just for luck, we named it the 'Coquette,' and thereby hangs a tale."[2] A. C. Kepler did not dispute that the well was named after his cousin's dream, but it appears that they selected the location not because of the dream but by tossing a hat in the air.

DORCIE CALHOUN'S DREAM

In 1936, Dorcie Calhoun dreamt of drilling into a huge deposit of natural gas on his Pennsylvania farm. In 1923, a natural gas company drilled on the farm but found only a bit of shallow gas. The small amount of gas was useless to the company, but Calhoun's father took over the well and piped the gas to the farmhouse. Dorcie always thought that the shallow gas was leaking up from a massive deposit. In 1936, he had a dream so vivid that it convinced Calhoun of its truth, although it would be another thirteen years before he acted on it.

In 1949, Calhoun told the editor of the *Renovo Daily Record* that he was going to prove his dream. Calhoun knew of an old drilling rig that he could buy for a modest amount and invited the editor to invest in the well. The editor wasn't about to rely on Calhoun's dream, so he ordered literature from the Pennsylvania Geological Survey, including a geological map of the area. He correctly noticed that the well was located over an anticline—very favorable for gas—but he missed the note that said that most segments of the anticline had been tested and found barren of gas.[3]

Calhoun raised $15,000 from twenty-six friends to form the Leidy Prospecting Company and drill into his dream. He bought the twenty-five-year-old wooden drill rig for $5,000 and started. The rig was junk. It broke down on the first day of drilling and continued to fall apart in different ways, to be patched back together by Calhoun using his long experience of coaxing more use out of worn farm equipment. He ran out of money after two months of drilling, with no sign of gas. Dorcie's friends stood by him. They raised more money and drilling resumed. The old cable-tool drill rig was designed to drill to 2,000 feet. In December 1949, thanks to Calhoun's improvisational genius, they were down 5,600 feet and seeing signs of gas in the Oriskany Sandstone. On January 8, 1950, they hit high-pressure gas that came up the borehole and blew the drilling cable out of the hole. They had a discovery.[4]

The great gas production from the Dorcie Calhoun #1 well brought swarms of landmen from other oil and gas companies, leasing all the farms surrounding Calhoun's. The Leidy Prospecting Company drilled the Dorcie Calhoun #2 later that year, and the #3, #4, and #5 in 1951. They had drilled up Calhoun's farm and had nowhere to go.[5]

In January 1951, two national magazines, *Newsweek* and *Collier's*, ran stories about Calhoun and his offbeat road to fortune. *Reader's Digest* reprinted the *Collier's* article. Wealth did not change Calhoun much. He still bought his clothes at the Army surplus store. The only noticeable differences were that he could afford an endless supply of ten-cent cigars, and he bought three new cars. But a little more than a year after his success story was featured in *Reader's Digest*, Dorcie Calhoun was broke.

Dorcie Calhoun had proven his dream, had his picture in national magazines, and became a local hero. The Leidy Prospecting Company had drilled up his farm, and his investors had received back $13.50 for every dollar invested. It was time to sell out to a gas company and put the money in the bank. But that did not happen.

Calhoun was swept into the optimism of the gas boom he had created. Finding gas seemed so easy that the Leidy Prospecting Company kept wildcatting, but all he drilled were dry holes. In addition, the original gas wells were quickly becoming exhausted. In June 1952, an unpaid oil field equipment company attached one of Calhoun's drilling rigs. Dorcie Calhoun had been rich for two years and six months before he went broke.[6]

Calhoun went back to wildcatting, but without success. In 1958, he was working construction. He died in 1975.[7]

DREAMIN' ALONG

In 1891, Casper Ketchner dreamed that a certain oil well then being drilled in Western Pennsylvania would be dry, but that a well drilled at another spot would flow a thousand barrels per day. He told his dream to the owners of the drilling well, and when the well turned out to be dry, they asked Ketchner to show them the spot where he had dreamed of the successful well. They drilled on the spot he showed them, and the well flowed a thousand barrels per day.[8]

Edith Day of Elmira, New York, dreamt of oil at a place near the Gaines oil field at Watrous, Pennsylvania. The vividness of the dream convinced her that it must be true, so she decided to drill the well herself if she had to. She denied that she was a spiritualist, but newspapers named her well the "spirit well." Day dreamt that oil would be at 4,000 feet, but the drill reached 4,100 in May 1906 with no oil. Drilling continued until the drill bit was lost in the hole at 4,842 feet, and the well was abandoned. It was said to be the deepest well in Pennsylvania.[9]

Thomas J. Jamieson's dead father often popped into his dreams to give his son advice. One night in 1917, his father stepped into T. J.'s dream to take him on a flying tour of what his father said were the oil fields of southern Alberta, Canada. An unidentified man joined the dream, showed the son where to drill, and uttered a bunch of unfamiliar names. When he looked at a map the next morning, Jamieson discovered that the unfamiliar names belonged to rural post offices.[10]

But how was T. J. Jamieson to pay for all the leasing and drilling? He spent the next few months in a quandary, until his father appeared in another dream and told him to form the Dreamfield Oil Company Ltd., which he did soon after. In December 1919 Jamieson advertised shares for sale in the Lethbridge *Herald*, and by the start of 1920 the new company's first well was boring downward. Dreamfield Oil drilled one dry hole after another, until the company ran out of money and stopped drilling in 1923.[11]

PETROLEUM SPIRITS

THE RELIGION OF SPIRITUALISM IN THE UNITED STATES STARTED in upstate New York in 1848 and spread quickly. Believers sat around tables in darkened rooms calling to dead relatives and beings from the spirit world. Spirits would write or speak through one of the sitters or even materialize in the darkened room, walk about, and touch those present. Mediums competed for clients by providing more dramatic séance phenomena and invented many tricks of the trade.

There were many spiritualists in northwest Pennsylvania when oil was discovered there in 1859, and it was inevitable that the new industry and the new religion would join forces.

JONATHAN WATSON: GOING BROKE LISTENING TO THE SPIRITS

Jonathan Watson was the first petroleum millionaire. He both gained and lost his fortune at least partly by relying on dowsing and information from spirits.

Edwin Drake bored for oil at Titusville in 1859 and finally found it, but he lacked the business sense to profit from the discovery. Jonathan Watson had the business savvy that Drake lacked. On the day after Drake struck oil, Jonathan Watson saddled up before dawn and made his "million-dollar ride" down Oil Creek to the farm owned by Hamilton McClintock, which had an oil seep. He arrived before noon, bought an oil lease from McClintock, then hurried further down the creek and bought another oil lease to the farm owned by John Rynd. The Age of Petroleum was one day old, and Jonathan Watson controlled oil rights to two large farms along the creek; within ten days he leased more farms. In addition to the leases, Watson jointly owned extensive tracts in the area through the lumber firm Brewer, Watson and Company. As Drake's driller Billy Smith noted, there were few wells along Oil Creek in which Jonathan Watson did not own an interest. By the time the country awoke to the importance of the Drake well, Jonathan Watson controlled large portions of the early oil territory.[1]

Watson was a spiritualist and asked the spirits where to drill wells. Spirits guided Watson as he drilled wells along Oil Creek, and his worth grew to an estimated four to five million dollars. Watson consulted spirit mediums and sometimes sensed the presence of spirits—often deceased friends of his—and sensed their advice.

His wife died in 1858, and in 1862, at age forty-two, Watson married eighteen-year-old Elizabeth Lowe, a child prodigy from upstate New York. Like Watson, his new wife was a devout spiritualist. She had had visits from the spirit world since childhood and was a popular lecturer on spiritualism since the age of fourteen. They moved to her hometown of Rochester, New York, in 1866, but returned to the oil region in 1868 where they built a mansion on the edge of Titusville.

By the late 1860s, the territory along Oil Creek had been drilled up, as had the nearby Pleasantville and Pithole oil fields. To find oil, Watson had to drill deeper and drill farther from known production. But when the drilling grew riskier, the spirits gave bad advice.

Jonathan Watson's fortunes crashed slowly in the 1870s. He and his wife did not dissipate his fortune by riotous living or conspicuous consumption. Their home at Titusville was a mansion, but not a palace. Watson contributed large sums to charities. He also cosigned notes to help friends in business deals, and some abused his generosity, leaving him to cover large debts. Contemporary accounts blame Watson's problems on dry holes and financing faithless friends.

The spirits that had guided him to wealth now lured him into bankruptcy. Watson invested heavily in drilling in the Millerstown and Parker oil fields in Armstrong County, Pennsylvania. The spirits had no better idea of where to drill than anybody else, and Watson drilled dry holes located by spirits until he went broke. In November 1876 creditors seized all his property and assets, except his home. Watson kept drilling, determined to recoup his fortune, but just dug himself in deeper. He may have been the unnamed "wealthy operator" a *Monongahela Valley Republican* article stated lost $100,000 drilling on dowsed locations in 1878.[2] He sold his mansion in 1881.[3]

The Watsons bought a fruit farm in California, and his wife moved there in 1880 with their daughter, supposedly to alleviate the daughter's hay fever. Jonathan Watson went west to join his wife and daughter in 1882 but found his reception by his wife less than congenial. He returned to Titusville later that year, after which his wife secured a legal separation. Freed from her marriage, Elizabeth Lowe Watson bloomed anew, and again became a popular lecturer in the United States and Australia, and a leading figure in California on spiritualism, votes for women, and temperance.[4]

Watson drilled a deep well at Titusville but gave it up as a dry hole in January 1885. In 1889, Watson rode the train along the Oil Creek past Watson Flats, where Drake had drilled the first oil well that started it all; the spirit of former business partner Daniel Crosley told him that this was the place to drill another well. Watson went to the site the next day, and two more spirits of former friends assured him that if he drilled there, he would get the biggest wells ever drilled in the world. Crosley's spirit had asked him to allow his son, James Crosley, to join in the well, but James Crosley told Watson that the oil along Oil Creek was exhausted, and the location "wasn't worth a damn." So Watson spent his last thousand dollars to buy the land, and then convinced some oil operators to drill it.[5]

The well on Watson Flats found oil at 450 feet, and when it made 400 barrels of oil on the first day, Watson's spirit advice seemed vindicated. The well made fewer than 150 barrels on the second day, however, and soon produced only water. Watson's spirit friends had handed him another failure, yet he took the brief oil production to be a sign of success.

Spirits advised him to buy some oil leases in Allen County, Kentucky, but soon after buying them, Watson learned that leases had already expired. It sounds as if the leaseholder bribed the spirit medium to set Watson up—not unusual.

Watson turned to oil dowsers. Two dowsers gave him two locations, both dry holes. The dowsers rechecked their locations and told Watson that the oil was actually beneath a creek. Undeterred, Watson diverted the creek, but only drilled more dry holes.

In 1893, oil men contributed $2,000 to support Jonathan Watson at a sanitarium at Clifton Springs, New York, where he was cared for by his daughter from his first marriage. The following summer he contracted gangrene in his toe, which quickly spread. Watson refused to eat and died in June 1894, at age seventy-five.

Before he died, Jonathan Watson asked a friend to tell his wife that he had forgiven her. Watson was enormously popular in Titusville, and in equal degree his estranged wife was not. In its obituary of Jonathan Watson, the *Titusville Herald* wrote that Lizzie Watson had "proved false to every vow," a bitter accusation to hurl in public print in 1894. But in California, her former life back East must have seemed so far away. Elizabeth Lowe Watson had a farm to run, speeches to make, causes to organize, and no time to look back.

Jonathan Watson, the first oil millionaire, had been ruined, in large part, by dowsing and messages from the spirit world.

ABRAHAM JAMES: THE SPIRITS SEARCH FOR OIL

In 1863, four people sat in a darkened room in Chicago, listening to the spirits. The spirits told one of them, Abraham James, to discover a great oil field beneath Chicago.

Abraham James had been a railroad conductor and gold prospector. His mother had a reputation as a seeress in the neighborhood of their Pennsylvania farm, and James said that as a child he saw dead people walking about his bedroom. His spiritualist talents blossomed when he associated with other spiritualists in Chicago. He sometimes seemed to speak Spanish, Italian, German, and "an Indian language."[6]

At a séance in the summer of 1863, the spirits told James that a great petroleum reservoir underlay the intersection of Chicago Avenue and Western Avenue, the highest point in Chicago. James went to the site, went into a trance, and selected the exact spot to drill.[7]

Chicago spiritualists bought forty acres of land surrounding the well site and began drilling in December 1863. When the well reached 229 feet without oil, the spirits again spoke through Abraham James, promising a great oil field. Drilling halted in late 1864 at a depth of 711 feet, when a strong stream of pure drinking water flowed out of the well. The spiritualists downplayed his failure to find oil, trumpeted the flow of water as a great success of spiritualism, and formed the Artesian Well Company to sell the water to Chicagoans.

Any well drilled that deep within forty miles would have hit the same artesian flow. James and his friends were just the first to go to the expense of drilling into the bedrock in the Chicago area. Oil exploration in the 1860s accidentally discovered a number of artesian aquifers in the Midwest because previous water wells had stopped at bedrock.

The Pleasantville Oil Discovery

James decided that his spirit friends might do better where oil had already been discovered, so he went to northwest Pennsylvania. He advertised in the *Titusville Herald* in July 1866 that he and the spirits were available for consultation.[8]

On Halloween 1866, Abraham James went in a carriage with three other men to examine some oil property. South of Pleasantville, James leapt from the carriage and climbed over the fence and onto the farm field. His companions followed him as he walked back and forth until he fell to the ground, stuck his finger into the earth, pushed a penny into the soil to mark the spot, closed his eyes and became rigid and apparently lifeless. When he woke, he told his companions that a mighty stream of oil flowed beneath the spot where he had put the penny. James and his partners leased the farm and awaited further word from the spirits.

In the meantime, he had his spirits find places to drill salt wells. In June 1867, James told the editor of the Cazenovia, New York, *Republican* that spirits had shown him three places near town to drill for salt brine, and that he would point them out for a hundred dollars and 5 percent of the proceeds.[9]

Salt was forgotten that same month after the spirits told him to drill the oil well at Pleasantville. Spiritualists raised funds, built a derrick, and began drilling in August 1867, to the hoots of the skeptics. The spirits counseled James while the drill bit broke its way down through the rock. On January 31, 1868, the Harmonial Oil Well No. 1 hit an oil-bearing sand at a depth of 832 feet. They drilled three feet into the sand and oil flowed to the surface, the rate gradually increasing to 135 barrels per day.

The discovery ignited an oil boom around Pleasantville, but despite James's precise location, marked with a penny stuck in the ground, the oil field was five miles wide and twenty miles long. Later drilling proved that he had placed his well toward the thin northern edge of the field, and wells further south produced more oil.[10] The Harmonial well produced oil in economic quantities for less than two years.

James sold his share of the first Harmonial well, then had his spirits choose four well locations nearby and sold partial interests in the Harmonial Oil Well No. 2 through No. 5.[11] The discovery made James famous as a "practical" spiritualist: one who harnessed the knowledge of the spirits for material gain. Guided by the wisdom of the spirits, how could he go wrong?

James also showed how easy it was to lose money by following spirit advice. In 1868 he bought a large block of leases at President, Pennsylvania. The area around President had boomed briefly a few years previous, when the first few wells were gushers, but later wells were uneconomic. James began drilling at President in April 1869, but no successful wells were recorded.[12]

The Deep Dry Hole

When his Indian spirit guide told him to drill a wildcat well on Blyson Run, in Clarion County, in 1871, James found investors and leased 3,000 acres around the location. If the spirits were right, he would discover a major extension of the Clarion field. The James and Crane well started drilling in February 1872. Typical oil-well depths at the time were a few hundred feet. In May 1872, as the well passed a thousand feet of depth, newspapers started referring to it as the "James well," indicating that his partner Crane had dropped out. He insisted that his spirits had not failed, that oil must be deeper.

Progress slowed as the well deepened. The drillers penetrated the first thousand feet in less than four months, but it took twenty months to drill the second thousand, and still no oil. James kept going deeper, until a reamer stuck in the hole, and he finally gave up after more than two years of drilling. His dry hole was 2,323 feet deep. His spirits had failed him, yet he optimistically spoke of forming another company to lease more land and drill more tests in the vicinity. But Abraham James had drilled up all his credibility and attracted no investors.

By some accounts Abraham James lost most of his profits from the Harmonial wells by drilling dry holes on spirit advice. But his profits from the Harmonial wells were said to be half a million dollars, and he didn't drill that many dry holes. After his discovery of the Pleasantville oil field, he became a generous contributor to spiritualist causes, and that may have drained his finances as much as his dry holes. By 1874 he was apparently no longer wealthy, but stories that he drilled himself into poverty appear to be exaggerated.

James walked away from his spectacular dry hole in 1874—at the time it was one of the deepest wells ever drilled for oil—and disappeared from newspaper headlines.

Reports of James's Death

One of his friends in New York received a letter in 1884 that said that Abraham James had died in Oregon. Major US newspapers carried the story, with a retrospective on James's oil career. But Abraham James was alive. The *Buffalo Morning Express* reported his death on November 29 but printed a retraction the following day. James had moved to Fredonia, in Western New York, where a *Morning Express* reporter spoke to his wife, who said that he was alive, that he had not been to Oregon in years, and he had recently moved to Florida to raise oranges, where she was about to join him. But few newspapers noticed the retraction, and news of his death kept popping up in smaller papers into 1885.[13]

James quit orange farming several years later and returned to Fredonia. He listed his profession as "oil operator," but he was better known locally as a physician, a Doctor of Vitapathy, a practice that the magazine *Medical World* called "[o]ne of the most reprehensible developments of medical humbuggery."[14]

Abraham James hit the jackpot—once—at Pleasantville in 1868. His spirits misled him at Chicago, President, and Blyson Run. He dabbled in oil locating for a quarter-century after his deep dry hole on Blyson Run, and if he ever made another big discovery, he would have shot back into newspaper headlines, but he never did. His continued obscurity after 1874 confirms his lack of oil-finding success. Like other spiritualists who drilled on spirit advice, Abraham James learned that the spirits of the dead were no better at choosing drilling locations than the living.

When Abraham James died in 1905, newspapers ignored the story. As far as the public was concerned, he had died more than twenty years earlier.

LIGHTER SPIRITS

One of Jonathan Watson's contemporaries nearly came to share his ruin. William Barnsdall was a shoemaker from England who had a shop in Titusville. He quit his shop as soon as Drake struck oil in 1859, drilled the second oil well in Pennsylvania, and became wealthy.[15]

The Barnsdall family were devout spiritualists. One of his daughters was a medium and began asking the spirits to recommend drilling locations. His first spirit well was a success, but all his daughter's later spirit locations were failures. Although he remained a spiritualist, William Barnsdall quit drilling spirit wells before they drained his fortune.[16]

According to William Wright, spirit medium Mrs. Winter ran a boarding-house in Virginia City, Nevada, home to the Comstock silver mines. The spirits told her that Mount Davidson, which rose on the west side of Virginia City, was a great hollow tank filled with petroleum. She showed the best location up on the slope of the mountain to tap the oil to Joe Griggs, an engineer at the Savage mill. Griggs spent eight days driving a tunnel to intersect the oil, until the spirits told him that his tunnel was close to the petroleum.

Griggs feared that breaking through to the oil would send a flood of burning oil down upon Virginia City. The spirits were consulted and recommended that a pipeline be installed to control the flow of oil. To lay the pipeline, Mrs. Winter invited some believers to form the Mount Davidson Oil Company, with herself at the head. All went well until Mrs. Winter levied a twenty-five dollar per share assessment, causing the investors to drop out.[17]

The above incident may never have happened. Its only source is William Wright, who wrote under the pen name Dan DeQuille for the Virginia City *Territorial Enterprise* alongside Samuel Clemens, where both amused themselves by slipping occasional tongue-in-cheek tall tales in with the news coverage. Wright included the Mount Davidson oil tank story in *The Big Bonanza*, his informal history of the Comstock Lode.[18]

Spirits told Belmont, New York, spirit medium James Robinson that oil underlay the Leilous farm one mile south of town. After an oil dowser confirmed the exact location, Belmont spiritualists formed the Spirit Oil Company. The well reached 3,510 feet in November 1896 and was abandoned as a dry hole.[19]

The Leilous farm somehow fascinated spiritualists. In 1899 two spirit mediums, Amanda Benson of Belmont and Lizzie Brewer of Syracuse, New York, received word from spirits that there were rich deposits of copper and gold at the Leilous farm. Spiritualists invested the money to sink a mine shaft but found no copper or gold.[20]

A spirit medium named Murphy told spiritualists at Bolivar, New York, that there was oil beneath the north bank of Barden Brook on the Patsy Garriety farm. The spiritualists tried to confirm the location by calling in Rev. Isaac Booram, who had some sort of oil detector. His oil indicator detected no oil, but the spiritualists believed the spirits. Drilling started in February 1920, and the well was abandoned at 1,776 feet after finding no oil.[21] As former spirit medium Thomas J. Minnock once stated: "When you get a good sucker, you always send him to the boneyard. Oh, well, the boneyard isn't a real cemetery, you know. That's our slang for the phony oil stock market."[22]

Spirit medium Alice B. Gardner ran séances at her Los Angeles home. In 1921, she often spoke in a trance about the Eliasville Pool Syndicate No. 1, which the spirits highly recommended as an investment. A federal deputy marshal arrested Gardner and her husband along with four others in 1923 for mail fraud. Alice Gardner was charged with using her séances to promote the Eliasville Syndicate, without telling her clients that the syndicate was paying her. The syndicate principal was found guilty and sentenced to two years of prison, but Alice Gardner and her husband were acquitted. She later became pastor of a spiritualist church in Pasadena, California.[23]

DECLINE OF SPIRIT WELLS

Spirits had early success with Jonathan Watson and Abraham James in the 1860s. But spiritualist oil drilling remained a novelty, and by the time Jonathan Watson failed in 1876, spirit drilling had largely discredited itself. The Geological Survey of Pennsylvania warned in 1885: "But the followers of spiritual or magical rites have averaged more failures and sunk more money for their friends and for themselves than any other class of operators in the country."[24]

Spiritualism in the United States declined after the 1870s, and drilling oil wells based on spirit advice declined as well. There were too many exposures of fraudulent spirit mediums for them to be taken seriously. To more and more people, the séance with a medium in a darkened room was just a joke from another era.[25]

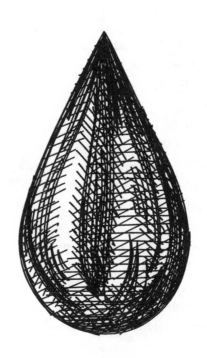

4

THE PSYCHICS

AS SPIRITUALISM DECLINED THROUGH THE TWENTIETH CENTURY, people still sought advice from clairvoyants, without the mystical trappings of the séance.

STELLA TELLER, WIZARDESS OF OIL

Stella Teller lived in Colorado, where her brother John Teller was a rich businessman, and her cousin Henry Teller was a United States senator. In October 1901, when she was forty-five, her brother had a court declare Stella insane, and the family sent her to a private asylum in Jacksonville, Illinois. The proof of insanity? She thought that she had the psychic power to find oil.

Stella watched how things worked at the asylum and waited. One night in July 1902, she stole a key and slipped out. She had made friends with the guard dogs patrolling the grounds and they let her pass without alerting the staff.

While the Teller family searched, Stella tried to get to Cleveland, Ohio, to offer her services to John D. Rockefeller to find oil in Egypt. She had only enough money to get as far as Valparaiso, Indiana, where a friend took her in. Her family learned her location the following month, when she sued her brother for $50,000.[1]

Dr. Sharpe from the asylum had no legal authority outside Illinois to force Teller back to the asylum. Sharpe spent more than two months lining up the legal authority, then he pounced. In October 1902, Dr. Sharpe and the Valparaiso town marshal forced their way into Teller's room at a boardinghouse, hustled her into a horse-drawn hack, and headed for the train station. Running close behind was Miss Teller's lawyer with a blank writ of *habeas corpus*. He got

Fig. 2. The doctor from the asylum and the town marshal forced their way into Stella Teller's room and hustled her onto a train headed back to the asylum. (*St. Louis Republic*, January 11, 1903.)

the writ signed and notarized, approved and stamped by a judge, and rounded up the county sheriff, who arrived at the train just as the conductor was about to signal it to pull out of the station. The officer boarded the train, served the writ, and freed Stella Teller. The drama of her family trying to kidnap and confine her in an insane asylum won a great deal of sympathy. At a court hearing the next morning, Stella was poised and well-spoken. After an all-day hearing, the judge declared her sane.[2]

Stella Teller started a twenty-year career as a psychic oil finder—or at least a psychic oil seeker. Her method was that at nine o'clock in the morning, she would lay out a map of the tract in question and, with pencil in hand, allow a spirit to direct her hand to mark the location of the oil, coal, or minerals. Teller told reporters that her family had declared her insane so that they could monopolize her psychic oil-locating for themselves.[3]

Stella Teller stayed in Valparaiso for a couple of years and searched for oil and gas in Ohio and Indiana. Then she moved to Texas and lived in Houston and Wichita Falls. By 1919, Teller was living in Wichita, Kansas, and billing herself as a geologist. Stella Teller was active in buying oil leases until 1922, after which she dropped from sight.[4]

THE BOY WITH X-RAY EYES

Guy Fenley was a phenomenon around Uvalde, Texas, because he could find water in that dry country. He was only a boy, but ranchers clamored for his help in locating their water wells. When he walked over the ground, he knew how deep the drill had to go to reach water—and he was said to be always successful.[5]

Fenley said that he could sense the depth of underground water since the age of two, but had just assumed that everyone else had the same ability. When he was nine or ten he mentioned the talent to his father, who tested his son by having him locate water on a ranch near Sanderson, Texas. The well found water close to the predicted depth, and Guy began to build a reputation. Although Fenley was uncomfortable about charging money, his father charged for each successful water well.[6]

Fenley's water sense was just a feeling, but it seemed unerring. In early 1901, when he was thirteen years old, newspaper headlines nationwide called him "the boy with the x-ray eyes."[7] This was just after the Spindletop gusher at Beaumont, Texas, established Texas as a major oil-producing state, so it was natural to bring the two Texas headlines together and turn Guy Fenley's talent from underground water to underground oil. Guy Fenley's father agreed to let his son look for oil.

Some businessmen formed the Uvalde Oil Company and leased a long strip of land a mile away from the new Spindletop oil field. They tested Fenley by burying a barrel of oil and a barrel of water in a field. Fenley walked the ground at night, as he said that his powers worked better in the dark, and found and identified both barrels. The Uvalde Oil Company then brought young Guy Fenley to Beaumont to find an oil bonanza. Fenley walked the tract alone, then told the Uvalde Oil Company men that their lease had oil at 1,200 feet. The drill went down on the Uvalde company lease in late 1901. In December, the press reported that the well hit a gusher of oil, and newspapers across the country carried the amazing story of the boy with x-ray eyes who could see oil in the ground. But the press report was mistaken, and the well had not found oil. Drilling continued until the company gave up the well as a dry hole in January 1902.[8]

Guy Fenley later told the story of the failed Uvalde Oil Company well. He said that he had gone to Beaumont with his father and older brother, who was a lawyer. His brother instructed him not to reveal the exact oil location before they had a signed contract with the company. Negotiations between his brother and the company broke down, and the Fenleys went back to Uvalde without revealing the exact location of the oil. The Uvalde company drilled anyway, but, according to Guy, drilled on the wrong part of the lease, and made a dry hole.[9]

Guy Fenley was uncomfortable using his God-given gift to make money, and so was not sorry to leave the oil business. His oil-sensing and water-sensing power disappeared after he returned from Beaumont. Nothing more was heard of Guy Fenley and his x-ray eyes. He later said, "I think the commercialization of my gift was what ruined it. It has been somewhat restored at times since I quit charging. But I've never again commercialized it."[10]

Her high school girlfriends warned Maggie Wilson to stay away from the "boy with the x-ray eyes" because he could see through her clothes, but Maggie had more sense than her friends. She allowed the tall blond boy to strike up a friendship with her; they later married. He went to college, then returned to the area and became a dairy farmer.

Guy Fenley never sought newspaper headlines and never wanted to use his water-finding ability to get rich. He enjoyed dairy farming. He could never refuse to help a neighbor who asked him to choose a place to drill a water well, but he never again charged a fee. Guy Fenley died in 1968.[11]

THE SEERESS OF CORSICANA, TEXAS

One of the most famous oil psychics was African American Annie Webb West, who was born in 1892 and grew up in the little town of Mexia, Texas, before it became an oil boomtown. She began foretelling the future as a child and became a professional clairvoyant, charging two dollars per reading. Her most famous psychic advice was the gusher location at Mexia that she recommended to Col. A. E. Humphreys. In gratitude, Humphreys gave her $8,500, which she used to buy a house in nearby Corsicana, Texas, her home for the rest of her life.

As she recalled, she told her brother-in-law not to sell his property in Mexia because there was oil beneath it. He sold the property anyway, so she recommended the property to Humphreys. She said that her location was the gusher oil well that set off the Mexia boom in 1920, though the timing is a bit off. She did not move to Corsicana until the mid-1920s, suggesting that it was some later oil well location, of the many that she recommended to Humphreys.

Born Annie Webb, she married eleven times to eight men—twice to one man, and three times to another.[12] She divorced ten times and was widowed once. Her first marriage was at age fourteen to farmer Arthur Buchanan. She divorced Buchanan, married and divorced Aaron Hobbs, then remarried Arthur Buchanan. She divorced Buchanan again, then married and divorced George W. Jackson.[13]

Her fourth husband was Rev. Clark West, eleven years her junior, whom she first married in 1931. Although they divorced twice, in 1936 and 1938, the attraction was such that she remarried West within a few months after each divorce. They sued one another for divorce again in 1940 but remained married until his death in 1942.

Annie set up each husband in business. Her marriage to James Lockett ended after only a few days when he asked her to put all her property in his name. The next morning, she told her manager to pack Lockett's belongings and take him to the bus station. Three more husbands followed; her last marriage was in 1956, followed by her final divorce in 1958.

While still in her twenties, Annie Webb Buchanan told wildcatter J. K. Hughes not to drill at the place he had planned and persuaded him to instead drill another location. He took her advice and the well came in a gusher. Then he went back and drilled the original location; as Annie foresaw, it was a dry hole. She was said to have a loyal clientele of wildcatters, including Howard

Fig. 3. Celebrated psychic oil finder Annie Webb is buried next to Rev. Clark West, whom she married three times. (Photo by the author.)

Hughes Sr., who would not drill without first consulting her. Despite her reputation as an oil finder, a court in Mexia fined Annie Buchanan in 1924 for fortune-telling.[14]

Wildcatter A. E. Humphreys was active and very successful in the Mexia oil boom of the early 1920s. He credited his success to his "triumvirate" of advisors: Almighty God, geologist Julius Fohs, and Annie Buchanan. Humphreys sold his oil wells in 1923 for thirty million dollars.[15]

Annie Buchanan became a legend in the oil business. Besides gushers at Mexia, she claimed to have correctly predicted the outcome of oil drilling in California, Oklahoma, East Texas, and the Corsicana oil field. Her fees rose to a premium five dollars per reading, but the greater part of her income was in gifts from grateful oil men. Besides her house in Corsicana, she owned homes in Houston and Springfield, Texas, and others. Some clients rewarded her with new automobiles—Pontiacs were her favorite.[16]

Annie Buchanan was a legend in Corsicana both for her psychic advice and for her charitable giving.[17] She recalled, "I built seven churches and a college for my colored peoples but I ain't never went to school a day in my life."[18] In September 1954 she was tried in Corsicana for vagrancy, on the charge that she was working as a fortune-teller. The hometown jury found her not guilty.[19]

Folklorist William Owens found her in 1956 with a waiting room crowded with clients black and white, some from far away, all of them ready to wait all day for a reading from a woman known to be as wise as she was illiterate. She read cards or palms, and sometimes healed with her hands. A sign advised that drinking was prohibited and that she would not give readings to women who smoked or wore pants. As Owens described the seeress: "She is a tall, remarkably well-built woman, marvelously well-preserved for the sixty-three years she claims. Her skin is a soft dark brown, her eyes dark, emotional, with a look that seems to pierce ordinary borders."[20]

Annie Webb Buchanan died at age seventy in 1962 and was buried alongside her favorite husband, Rev. Clark West.

EDGAR CAYCE'S MOTHER POOL OF OIL

No one has ever pumped a barrel of oil from San Saba County, Texas, but not for lack of trying. Many dry holes have been drilled in the county, and it is likely that more dry holes will be drilled, for San Saba County has a peculiar attraction for oil drilling: a hundred years ago, "the sleeping prophet" Edgar Cayce saw in a trance that the county contains the biggest oil field in Texas, a field he called the "Mother Pool."[21]

Edgar Cayce would go into a sleeplike trance and answer questions on everything from a person's health to the mythical continent of Atlantis. Cayce in a trance (trance-Cayce) assumed a more forceful personality than Cayce awake (awake-Cayce).

Cayce located his Mother Pool deep in the heart of Central Texas and drilled two dry holes there himself. Others have since drilled more holes, lured by Cayce's visions, but accomplished only more dry holes. The many books written about Edgar Cayce portray the dry hole at San Saba as the only time he failed—unique among his many successful predictions. America's most famous psychic tried to predict outcomes for more than a hundred wells across the United States. As an oil finder, he was nearly a complete bust.

Edgar Cayce is known for his readings on reincarnation, and about the lost continents of Atlantis, Lemuria, Mu, and others. Cayce predicted that in certain years of the late twentieth century, great land masses would rise up out of the water while others would disappear into the sea.[22]

Cayce's early fame was based on his health diagnoses and treatments, though doctors might doubt their value. He recommended carrying a piece of steel in the pants pocket to ward off colds and suggested eating an almond a day as a sure cancer preventative. For lung cancer, he prescribed inhaling apple brandy fumes from a charred barrel. His treatment for breast cancer was cocoa butter, or sometimes carbon ash with iodine ointment, massaged into the breast. Trance-Cayce said that smoking three to eight cigarettes per day was good as it supplied that healthy stimulant, nicotine. He sometimes prescribed "radium water."[23] Cayce wanted to use his psychic power to raise money for a hospital where patients would be treated in accordance with his trance-readings.

The Wildcat at Comyn, Texas

In 1919, the Sam Davis Oil Company exhausted its capital drilling a well outside the crossroads settlement of Comyn, Texas. It was a wildcat, thirteen miles from the nearest oil field, and there was no good reason to choose that particular spot to drill. They stopped drilling at 3,500 feet, a dry hole, despite a show of oil at 3,400 feet. But not all the investors gave up. Day Matt Thrash, the business manager of the Cleburne, Texas, *Morning Review*, sent Cayce information on the well, asking him for help.

Oil discoveries in Texas were making headlines and creating fortunes for the bold and the lucky, so it seemed like child's play for a seer like Cayce.[24] He was forty-two years old and was supporting his family with his photography studio in Selma, Alabama. His trance-readings were a sideline that paid little. Cayce went into a trance and described in detail the Sam Davis borehole—although it is unclear how much he was repeating information received from Thrash. Cayce recommended that they clean out obstructions in the borehole and "shoot" the well with nitroglycerin. The reading impressed the investors enough to invite Cayce to Texas.

In July 1919, Edgar Cayce and his friend David Kahn stepped off the train in Cleburne, Texas, to meet Matt Thrash and other Sam Davis investors who believed in Cayce enough to reenter the well. But not all investors agreed, and to fund the gap Cayce's friend David Kahn found out-of-state investors, and Kahn became vice president of Sam Davis Oil.

While the Sam Davis Oil Company was reorganizing, the Lucky Boy Oil Company asked Cayce for readings on wells it was drilling at Desdemona, fourteen miles north of Comyn. According to Joseph Long, one of the Lucky

Fig. 4. Central Texas, showing areas of Cayce Petroleum activity.

Boy Company officers, the "Lucky Boy No. 2" was thought to be a dry hole at 3,500 feet deep, but Cayce recommended that they plug back to about 3,000 feet and shoot the well with nitroglycerin. They did this and began producing 600 barrels of oil per day.[25]

The story seems exaggerated, however. The well that fits the description is the Lucky Boy No. 2 Ellison. The well was drilled to 3,500 feet, and in July 1920 was reportedly making 640 barrels per day after shooting with forty quarts of nitroglycerin. Several weeks earlier, however, the well had been drilling at 3,354 feet with a borehole full of oil—which would not appear to be a dry hole. But taking Mr. Long at his word, Cayce saved the well.

In addition to his successful prediction on the Lucky Boy well at Desdemona, Cayce made two other successful oil predictions in early 1920.

He predicted that a well being drilled in Washington County, Oklahoma, would find oil, and it did. Cayce said that a well being drilled near Bowling Green, Kentucky, would be a dry hole, which it was.[26] Cayce was three for three, and his oil predictions seemed infallible. Flushed with success, Cayce thought that making his fortune in oil was a sure thing. Kahn exulted, "No man in the world can stop us now."[27]

His biographers portray Edgar Cayce as a sensation in Texas, celebrated in front-page newspaper headlines. But a search of old Texas newspapers found no headlines about Cayce's predictions. The Texas oil industry was awash with psychics, dowsers, oil smellers, and inventors of homemade oil detectors, and Cayce did not stand out. Wells were being drilled at Desdemona and nearby Ranger by the hundreds, and many made thousands of barrels per day. That the Lucky Boy No. 2 Ellison at Desdemona started at 640 barrels per day was not news and surprised no one.

The future appeared limitless for Edgar Cayce, but this would be the dizzy high point of his oil-finding career. From then on, as far as is documented, his many attempts at oil-finding spanning the next twenty-plus years would be an unbroken string of failures. Edgar Cayce would never find another drop of oil.

The Sam Davis Oil Company reentered their abandoned well in February 1920. Cayce soon became convinced that he was the victim of sabotage. Machinery broke down and a rival oil company supposedly secured a lease that would take effect after the Sam Davis lease expired. Driller Cecil Ringle ran off at gunpoint a hobo who he thought was a disguised saboteur. What should have taken a few weeks stretched into months, until the company once more ran out of money.

Despite the supposed sabotage, no rival companies rushed in to drill after the Sam Davis Company walked away. Cayce's follower Cecil Ringle returned the following year, obtained a lease, and once more reentered the well, and once more was unable to extract oil. The failure of the Sam Davis well was an example no one wanted to repeat, and Comyn remains today a little crossroads with no oil wells for miles around.[28]

Comyn, Texas: 1 unsuccessful recompletion

The failure at Comyn convinced Cayce that he needed his own oil company, so in August 1920, he incorporated the Edgar Cayce Petroleum Company and engaged a New York stockbroker to sell shares, but the broker found that investors were unwilling to back a psychic. The Edgar Cayce Petroleum Company raised $50,000.[29]

The Not-So Near Miss at Luling Oil Field

"Neither did the operations of the Cayce Petroleum Company prove to be worthwhile. They never got production. I am still wondering why."

—Edgar Cayce[30]

Cayce biographers write that had he but drilled a well before his six-month lease expired, he would have discovered the Luling oil field and become rich. Some recent stories give Cayce the credit for advising Edgar B. Davis where to drill the Luling field discovery well, but Cayce never came close to discovering the field, and there is no record that it was found on his advice.

Cayce's decision to explore at Luling may have been suggested by Joe Rush, an oil driller he met in Desdemona. Rush had worked for Morris Rayor, who spent years drilling dry holes near Luling under the guidance of a psychic from Detroit.

Oil, gas, and salt water leaking to the surface near Luling had inspired prospectors to drill for oil—unsuccessfully—since 1902. When Cayce arrived in 1920, drill rigs were probing the Luling area for oil. In July 1919, geologist Vern Woolsey discovered a fault cutting the rocks in the riverbed of the San Marcos River. His mind's eye followed the fault thousands of feet down to a large structure where oil could be trapped. Woolsey accurately predicted that oil was trapped against a fault at Luling, and he recommended a location to his employer, Boston millionaire Edgar B. Davis. But zeroing in on the actual oil reservoir took a few years and a few dry holes.

According to Cayce lore, he leased 5,000 acres in Gonzales County. Cayce said that beneath the property there was "a great salt dome loaded with the richest oil in Texas," at a depth of only 300 feet. But the leases gave Cayce only six months to start drilling—and Cayce didn't have the money.[31] Despite gushing hometown publicity that people clamored to invest in Cayce Petroleum, the company was short of cash from the start. The company arranged to buy drilling equipment but lacked the funds to take possession. By the end of March 1921, the company had lumber for an oil derrick stacked at a drilling location near Smiley, Texas. Cayce couldn't raise the money soon enough, and his leases expired. Cayce's drill site at Smiley, which he never drilled, could not have discovered the Luling field because no part of that field is in Gonzales County, and Smiley is more than thirty miles from the Luling oil field.

Cayce came to Luling briefly then left, but Edgar Davis stayed on and drilled six dry holes in 1921 and 1922. Davis bet his fortune that geologist Vern Woolsey was right. Davis was on the edge of bankruptcy and was about to default on the loans and lose the field to his creditors when his next well hit a gusher of oil and secured his fortune.

Edgar Davis was famously generous. He gave money away by the millions, and he certainly would have rewarded Cayce handsomely had Cayce contributed to his success, but Davis gave Cayce nothing. The two Edgars never met, and, as far as we know, Davis never asked for or received Cayce's advice on where to drill oil wells.[32] However, Edgar Davis had a mystical inclination, and years later he had Cayce give him three readings on past lives.

Cayce predicted that the oil at Luling would be associated with a salt dome, but there is no salt dome within fifty miles of the Luling field. Davis followed geologist Vern Woolsey's idea that oil would be trapped against a fault, and Woolsey—not Cayce—was correct.

By March 1921, Cayce realized that the company could afford to drill only one well. He chose to drill in San Saba County, Texas.

The Mother Pool

"... at twenty-six [2,600 feet] we find the largest body of oil in the whole state, or country in this part of the sphere. ... The production in this we will find will be 70,000 barrels of 42 gravity oil per day ..."
—Edgar Cayce, on Rocky Pasture, San Saba County, Texas,
December 29, 1923

Cayce predicted a great oil field in San Saba County, a blank spot in the map of Texas oil fields. Oil drillers had poked only dry holes in San Saba County, but Cayce predicted that his well would find the "Mother Pool" at 2,600 feet and would produce oil at 8,000 barrels per day.

In September 1920, Cayce signed a lease with the Live Oak Oil Association for 4,000 acres seven miles northeast of San Saba, at Rocky Pasture on the Julia

Moore farm. If Cayce discovered oil, he would earn the right to produce oil from the 640-acre tract (one square mile) surrounding Rocky Pasture, plus the option to drill a checkerboard of 160-acre tracts covering half the surrounding area, with Live Oak Oil keeping the other half of the checkerboard.[33]

In July 1921, Cayce Petroleum began drilling the Live Oak #1, also called the Rocky Pasture #1, or the Julia Moore #1. Drilling was slow, and the hole inched down through the summer and fall. Cayce wrote that the Rocky Pasture well had "been expertly sabotaged a dozen times."[34] The crew began carrying pistols and guarding the well around the clock—but problems blamed on sabotage persisted.[35]

In November 1921, the well reached a depth of 1,560 feet, and Cayce decided to produce oil from the zone from 1,100 to 1,270 feet. Following common practice, sixty gallons of nitroglycerin were lowered into the hole and exploded. The drillers bailed out the shattered rock to allow the oil to flow, but the zone did not yield oil. Cayce started drilling once more but gave up at 1,642 feet deep, still a thousand feet from hitting the supposed Mother Pool.

A Cayce biographer wrote that two sets of tools were lost, presumably stolen. Someone supposedly dropped a piece of a tombstone down the well to convince Cayce that the hole had reached granite bedrock and there was no point in drilling further. But it is possible that these incidents have been misinterpreted. In cable-tool drilling, the method used at the well, "losing the tool" was a common occurrence and meant that the cable broke or the bit became detached and had to be fished out of the hole. In oil-field parlance, tombstone is a layer of hard, impermeable rock. It is possible that someone unfamiliar with drilling slang took the phrase literally. Most of Cayce's readings from this era were not preserved, and memories had twenty years to grow before the story was first put into print. The proof that Cayce was not the victim of sabotage is that no one rushed in to lease and drill his sites after Cayce's failure. Yet Cayce, his associates, and his biographers promoted the sabotage theory.

Cayce at first planned to move the equipment a few feet and redrill the well but, for reasons unknown, he hauled the drilling equipment four miles south, reassembled it on Polecat Ridge, and drilled the Munsell well. Cayce ran out of money and abandoned the Munsell well at a depth of 798 feet.

San Saba County, Texas: 2 dry holes

For the rest of his career, Cayce continued trying to predict results of oil-well drilling, at more than a hundred locations in seventeen states. His oil predictions were very optimistic and uniformly bad. The number of dry holes drilled on Edgar Cayce's advice is difficult to determine. Records of wells from that era are scattered and incomplete, and well locations are often only vaguely described. Many Cayce oil locations were apparently never drilled.

Cayce tried to arrange a share of any oil profits from his predictions for himself. In 1924, when a driller wrote to him about oil-well locations in Oklahoma, Cayce demanded to know what portion of the profits he would assign to the Cayce Institute. When the driller answered vaguely that it would depend on a number of factors, but that he would be "very fair in case I got a good well," Cayce refused to make the reading.[36]

Madison Wyrick's Dry Hole from Hell

Madison Wyrick was a Western Union executive in Chicago who was convinced that Cayce's advice had helped his diabetes. In 1924, Wyrick asked for a reading on the oil possibilities of his farm in Bosque County, Texas. Cayce told Wyrick to drill twenty paces southeast from a tree stump near an old house, where a well costing $18,000–$23,000 would produce between 3,000 and 3,500 barrels per day from a depth of 2,929 to 3,300 feet. Wyrick wrote back that there was such a house and stump, but that twenty paces southeast would put the well off the property. Cayce corrected himself and specified a location twenty paces northwest of the stump.

Wyrick incorporated and sold shares in the Telegram Oil Company and started drilling in September 1924. The hole went down easily past the depth where Cayce had predicted oil. But not to worry, for Cayce now said that the oil would be found at 3,635 feet. By December 9, the drillers were down to 3,660 feet with no sign of oil, but Cayce assured Wyrick that the drillers would find the oil within the next few days.

Again and again, Cayce's trance readings predicted oil just a bit deeper than the drill bit, then the hole passed that depth without finding oil, and Cayce once more changed his prediction, and said that the oil was a bit deeper still. When the hole reached 3,921 feet, Cayce said that oil would be found at 4,040 feet. When the drill penetrated to 4,242 feet, the oil was just a bit deeper. Deeper and deeper, chasing the oil bonanza, which kept sinking just below their reach. Cayce, from his trance, explained that the oil had moved since his first reading, due to underground "slides." This was nonsense, of course, but trance-Cayce always spoke with such commanding confidence that Wyrick kept drilling.

In a remarkable series of seventy trance readings, sometimes one every few days, for months, for years, trance-Cayce kept assuring Wyrick that all he needed to do was drill a few more feet, a few more days, and the great treasure would be his. Even as each prediction proved wrong, Cayce admonished Wyrick to "Be diligent," "keep on keeping on," "patience, persistence, and consistent effort will bring results," and "Be not faint-hearted, nor give up."[37] At a drill depth of 4,433, the oil was down at 4,568. By the time the drill reached 4,571, the oil had retreated to 4,622 feet. Deeper and deeper. In August 1925, with the hole down to 5,900 feet, the rig burned down. But still believing in Cayce, Wyrick rebuilt the rig and drilled on.[38] By March 1927, after two and a half years of drilling, they were at 6,105 feet when they shot a round of nitroglycerin to dislodge an obstruction in the hole, but collapsed the casing, making it impossible to drill further.

The well was a complete $100,000 failure. Yet Cayce's final trance reading on the well, in April 1927, was upbeat, and he advised that the company drill another hole 150 feet away, and counseled in his garbled way: "We would not give up the project for oil and gas will be produced in this territory, at much less depth than was drilled in this hole, as has been thoroughly illustrated or demonstrated by that has happened in these operations."[39]

Bosque County, Texas: 1 dry hole

Trance-Cayce was certain, and the more he was proved wrong, the more strident his certainty became. Wyrick chose trance-Cayce's self-certainty instead of the undeniable fact that trance-Cayce was wrong again and again.

Although Madison Wyrick still had faith in Cayce's medical readings, he did not bother to consult Cayce when he promoted a wildcat well in Crockett County, Texas. One of Cayce's acolytes asked for advice on whether or not to invest in Wyrick's Crockett County wildcat, and Cayce assured him that the well would find oil at 1,200 feet and would flow to the surface at 10,000 barrels per day. The well went down to 1,400 feet and found only water.

Crockett County, Texas: 1 dry hole

Cayce did a reading for land in Yuma County, Arizona. Cayce said that oil could be found at a depth of only 485 feet, but it was "migratory oil," which caused it to flow only from 11 p.m. to 11 a.m. (Mountain Standard Time) each day. No oil has ever been found in Yuma County, and no oil well has ever been drilled in which the oil flowed only between certain hours. But, as far as is known, none has ever been drilled on the spot in Arizona specified by Cayce.[40]

Another Mother Pool in Florida

". . . which will eventually become one, or the largest oil center in the country, surpassing even the Tampico fields, from which this is, as seen, a portion of the field, a structural continuation."

—Edgar Cayce, on northwest Dade County, Florida[41]

In 1920, Cayce did a trance reading for some Miami businessmen in which he saw a huge oil field in South Florida, a continuation of the rich oil fields of Tampico, Mexico, for which he predicted flows of ten thousand to forty thousand barrels per day from depths of 1,620 to 1,900 feet. By 1926, the businessmen had leased fifty square miles, organized the Miami Oil and Natural Gas Company, and brought a driller from Texas. By September 1926, the hole was down to 2,109 feet, with no oil. Cayce told them to keep drilling.

By January 1927, they were at 2,578 feet, and a great deal of water was flowing out of the borehole. Trance-Cayce gave some mumbo-jumbo about the water pressure being caused by oil and gas. He told them that they had drilled through the oil reservoirs, and that they had to stop the flowing water to produce the oil. He advised them to drill more wells nearby, to drain off the water, but it was useless advice, because they had drilled into the Floridan aquifer, whose artesian flow had nothing to do with oil and gas. The investors gave up and abandoned the well in 1932, at a depth of 4,560 feet. Since then, enough oil tests and deepwater wells have been drilled in the area to know that Cayce's huge shallow oil field in South Florida was purely imaginary.

Dade County, Florida: 1 dry hole

The Return of the Mother Pool

"Here we should find exceedingly large production—the largest as will have been encountered in Texas, see? For this should reach that proportion of 50 to 75 thousand barrels per day, see?"

—Edgar Cayce, on the Wilmott well, San Saba County, Texas
January 30, 1927[42]

In 1924, Cayce began giving oil readings to oil man Curtis Wilmott, whom he had met on a train. Cayce predicted oil at four locations in Oklahoma, and Wilmott drilled at least two—both dry holes. But Wilmott still believed in Cayce's oil predictions.

Cayce urged Wilmott to drill the Rocky Pasture location in San Saba County, so in 1926, Wilmott leased the Julia Moore Ranch at Rocky Pasture. Despite Cayce's statements that his well had been sabotaged by rivals, no one else had drilled the property since Cayce left four years previous, and Wilmott had no trouble leasing the tract.

Cayce gave Wilmott at least nine readings between April 1926 and January 1929, all assuring him that his well would drill into the Mother Pool at about 2,600 feet. In September 1926, Wilmott started drilling his Wilmott #1 well near Cayce's Live Oak #1 dry hole. Wilmott abandoned his first attempt after losing a string of tools down the hole. In April 1927 he moved the drill rig fifty feet and began drilling the Wilmott #2.

In December 1927, Cayce predicted that Wilmott would successfully complete the well in February of the following year. But by March 1928 Wilmott was down to 3,090 feet, well beyond the 2,600-foot depth at which Cayce had predicted oil, yet Cayce continued to predict that Wilmott would find oil at about 2,600 feet. Cayce's last reading for Wilmott's Rocky Pasture well was in January 1929, when he assured Wilmott that his well would find the Mother Pool before it reached 2,774 feet, and that the well would be a success. Curtis Wilmott stopped writing, and his whereabouts were a mystery. Years later, Cayce received a letter:

"I am here at Dr White's Sanitarium, taking treatments for alcoholic problems. I have been drinking pretty hard since the well in Rocky Pasture was dry."

—Curtis Wilmott, March 27, 1935[43]

Wilmott had fully tested trance-Cayce's prediction that a large oil field was 2,600 feet below the surface at Rocky Pasture. That should have killed the idea, but it didn't.

Oklahoma: 2 dry holes

San Saba County, Texas: 2 dry holes

Cecil Ringle's Twenty-Five Years of Dry Holes

Perhaps the person most abused by his faith in Edgar Cayce's oil predictions was driller Cecil Ringle. For more than twenty years until Cayce's death in 1945, and for several years afterward, Edgar Cayce's oil predictions kept Ringle impoverished.

Cecil Ringle was the driller on Cayce's Sam Davis well in 1920. Ringle was convinced of Cayce's psychic power, and after Cayce abandoned the Sam Davis well, Ringle leased the oil rights.[44] He returned to the Sam Davis well in 1922 and spent a couple of months trying to complete it as an oil well, but he could not extract oil from the well.

Ringle mostly drilled wells for other people, but whenever he saved up some money, he went wildcatting on his own, and with Cayce telling him where to drill, success seemed a certainty. Instead, Cayce's advice led him to twenty-five years of failure.

In 1924, he wrote Cayce for a reading about oil in Wichita County, Texas. Trance-Cayce told him exactly where to drill to find oil at a depth of 492 feet. Ringle drilled to 700 feet, without the smallest indication of oil.[45]

In 1925, while Ringle was drilling on contract in Northern New Mexico, Cayce advised him that the well would find a small oil reservoir at about 770 feet. Several months later Ringle wrote back that the well was at 950 feet with no oil—another Cayce failure.[46] But Cayce had correctly predicted that the hole would have problems controlling water, and that was enough to sustain Ringle's high regard for Cayce.

In 1929 Cecil Ringle moved to Ohio to drill for oil. The name of Ringle's company, Mystic Oil, as well as his warm friendship with Cayce, suggest that he relied on trance-Cayce's advice. Spanning the next few years, Ringle drilled an unbroken string of five dry holes in Ohio, but it is not known if any of the locations were endorsed by Edgar Cayce.[47]

Ringle left Ohio in 1934 to drill in New Mexico. He asked Cayce if there was any oil under the Tome Grant near Belen, New Mexico. No oil had been found within a hundred miles of the tract, so it was risky wildcat country, but Cayce assured Ringle that the tract had excellent oil zones at 2,900–3,000 feet, 3,200 feet, and 3,800 feet.[48] Cecil Ringle and his wife spent all the money they had to lease the entire 42,000-acre tract.

In 1934 and 1935, Ringle drilled three dry holes on Cayce's advice. He then found an oil man named George Grober with some money and an oil-finding gizmo supposed to be 97 percent accurate. The Grober #1 Tome started drilling in 1937. In September 1939, Ringle wrote Cayce that he and his wife were "flat broke" and had put their house in the name of a friend as protection against creditors. Ringle held on to a hundred thousand acres of leasehold—about 160 square miles—and the Grober #1 Tome well was down to 3,047 feet, and he was still hopeful. Ringle abandoned the Grober well in 1940 at 3,978 feet, without finding oil.

All four wells were dry holes, but after Cayce died, Ringle took another lease on the Tome Grant and drilled three more dry holes in 1947.[49] After drilling seven dry holes on Cayce's advice, Ringle wanted to drill still more, but the lease expired.

Losing the Tome Grant lease was the best thing to happen to Cecil Ringle. With his lease gone and Edgar Cayce no longer around to give him bad advice, he gave up the oil business and became a highly successful manufacturer and real estate developer in Albuquerque. Despite all the dry holes, Cecil Ringle never lost faith in Edgar Cayce, and when he died in 1960, he left one-fifth of his estate to Edgar Cayce's foundation.

Comanche County, Texas: 1 failed re-entry

Wichita County, Texas: 1 dry hole

Rio Arriba County, New Mexico: 1 dry hole

Tome Grant, New Mexico: 4 dry holes

After Cayce died: 3 more dry holes at the Tome Grant, New Mexico

A Second East Texas Field

In May 1940, trance-Cayce was asked about an oil test being drilled in Smith County, Texas. Cayce predicted that a great oil pool was beneath the entire property and said that it would produce a thousand barrels of oil per day from a depth of 4,079 to 4,082 feet.[50]

The hole drilled way past the predicted oil depth, but in session after session, trance-Cayce insisted that the hole be drilled deeper until oil was found. Cayce's wife Gertrude, who posed questions to Cayce while he was in the trance, started asking why the readings were inaccurate. Trance-Cayce became testy and spoke in garbled sentences, insisting that the readings were accurate. If pressed, he might suddenly end the reading with his phrase: "We are through for the present." During a frustrating reading in May 1940, Cayce again insisted that the well drill deeper. He emphasized: "Do not allow anything to interfere with there being brought production in these present operations." Gertrude challenged her husband: "Is it advisable and wise for Edgar Cayce to give information of this nature under any circumstances?" He answered, "As we have indicated in the above, it must be analyzed in the minds of the individuals seeking. As we have given, we would not advise such."[51]

In a reading in November 1940, when his wife again quizzed him about the well, trance-Cayce again insisted that the well contained oil, and that the rock cores from the well contained substances valuable for metal fabrication and for the manufacture of explosives and should be tested for those properties. But when Gertrude asked him to specify exactly what tests should be run, he petulantly repeated that they should be tested for use in preparation of metals and manufacture of explosives. Trying another tack, she asked him who could do such testing, and trance-Cayce snapped back: "One who is capable of doing so!" When his wife again pressed him about inaccuracies in previous readings, trance-Cayce ranted: "Come! Harken ye children of men! Bow thine heads. Ye sons of men!" Then he announced, "We are through for the present."[52] The location was a dry hole.

Two years later, Cayce gave readings on two more wells in Smith County. Cayce predicted that they would both produce oil. They were both dry holes.[53]

Smith County, Texas: 3 dry holes

Some More Oil Predictions (All Wrong)

Many of Cayce's well locations were described too vaguely to track down. Many other predictions were never drilled. Finding the results of wells drilled in the 1920s and 1930s can be difficult, but the following are more results of Cayce oil predictions:

1920, Comanche County, Texas. Cayce predicted oil. Result: dry hole.[54]

1920, Warren County, Kentucky. Cayce predicted that two wells would have no commercial production. Result: two wells each pumping twenty-five barrels of oil per day. These were development wells, and while not spectacular, they appear to have been commercial.[55]

1924, Okmulgee County, Oklahoma. Cayce predicted oil. Result: dry hole.[56]

1928, Coleman County, Texas. Cayce predicted oil. Result: dry hole.[57]

1928, Morehouse Parish, Louisiana. Cayce predicted that the well would be "an especially well paying proposition." Result: dry hole.[58]

1928, Crockett County, Texas. Cayce predicted oil. Result: dry hole.[59]

1932, Blunt County, Alabama. Cayce predicted oil. Result: 3 dry holes.[60]

1939, Pontotoc County, Oklahoma. Cayce predicted oil. Result: dry hole.[61]

Cayce moved to Virginia Beach, Virginia, and started the Association for Research and Enlightenment (ARE), a group interested in spiritual development. Thomas Sugrue's biography of Cayce, *There Is a River*, became a bestseller in 1942, and by the time Edgar Cayce died in 1945, he was America's most famous psychic. His sons and a few close associates strenuously kept alive the interest in Edgar Cayce, and membership in the ARE grew. Today the ARE boasts 39,000 members in seventy countries and operates Atlantic University, a postgraduate institution for spiritual studies, with twenty-five faculty and more than one hundred students.

The Ongoing San Saba Follies
Cayce died in 1945, still convinced that the Mother Pool was there waiting to be found.

Rudolph Johnson was an attorney in Dallas who in 1950 decided to show the world that Cayce was a true prophet by finding the Mother Pool. Johnson went to San Saba and found Wilmott's dry hole but decided that Wilmott had drilled the wrong spot. Cayce had said to go thirty paces north from the junction of two creeks, but Wilmott had apparently paced off in a direction sixty degrees east of north. The discrepancy in location cannot have been much more than a hundred feet. Johnson collected $40,000 from Cayce fans, leased 18,500 acres, and drilled down to 3,520 feet without finding oil. Johnson wanted to drill deeper, but ran out of money.[62] Johnson's well was the second to fully test Cayce's Mother Pool prediction, and the second to prove it wrong.

A company named the Cayce Petroleum Corp., Inc., drilled a well at Rocky Pasture in 1981 to a depth of 3,200 feet. It was a dry hole.

Petroleum engineer David Magnum joined with oil operator Jerry Conser to find Cayce's Mother Pool. They made an electrotelluric survey of the area and selected a site a mile from the original Cayce location. In 2002 they drilled the Horne #1, a dry hole to a depth of 2,400 feet.

David Magnum formed San Saba Exploration, LLC, to drill another well. A write-up for the project warned: "If you do not believe that Edgar Cayce was credible, then read no farther."[63] But the well was not drilled. In 2011, Cayce Operating Company permitted the Rocky Pasture #1, but it was never drilled. In 2014, the Cayce Energy Company was raising money to drill another well in search of Cayce's Mother Pool.

San Saba County: 3 dry holes at Rocky Pasture drilled after Cayce died

There is much of the little boy in the Cayce readings: love of adventure and drama, past lives as famous people, palace intrigue in ancient Egypt, and great land masses suddenly sinking into the ocean. Cayce's get-rich-quick schemes included oil gushers, lost gold mines, pirate treasure, and a perpetual-motion machine. These are all the stuff of a little boy's daydreams, but like most such daydreams, none of them succeeded. Yet trance-Cayce's boyish optimism survived every failure.

By the time Cayce died in 1945, it was obvious that no one should drill an oil well on his say-so, and the dry holes drilled since then have confirmed the obvious. If his followers knew of Edgar Cayce's long list of dry holes, maybe they would quit drilling dry holes in San Saba County. Maybe.

H. L. HUNT: ARE YOU LUCKY?

"Finding oil is all luck. Some people are born lucky and are lucky all their lives."

—H. L. Hunt[64]

Haroldson Lafayette Hunt Jr. believed in drilling for oil using luck, dreams, psychics, hunches, and four-leaf clovers. This sounds like a recipe for financial ruin, but H. L. Hunt was one of the richest men in the United States. He was a gambler and real estate speculator, novelist and songwriter, political philosopher and bigamist. He was a billionaire who carried his lunch to work in a brown paper bag. There was never anyone else quite like H. L. Hunt.

Hunt grew up in Illinois and went south as a young man to speculate in cotton land in Arkansas, with indifferent success. He had more success when he opened a gambling room in the oil boomtown of El Dorado, Arkansas. Not only did the gambling pay well from the house cut, but Hunt won a good deal of money by playing poker himself. He calculated the odds better and faster than anyone else at the table and called himself "the best poker player in the world."[65] His gambling drew him into the oil business when he won an oil lease in a poker game. He had to borrow money to drill a well on his lease but the well hit a gusher. He repeated his success on some other oil leases.[66] Hunt became a "lease hound," buying and selling oil leases.

When Florida real estate boomed in the 1920s, Hunt was in the thick of it, and made a lot of money—on paper—until the bubble burst and he barely got

Fig. 5. The iconic photograph of the Daisy Bradford #3, taken September 30, 1930. (Photo courtesy of East Texas Oil Museum, Kilgore, Texas.)

out of Florida with his shirt. He also came out of Florida with his second wife, never mind that he was still married to his first. Wife #1, Lyda Bunker Hunt, was raising their children in El Dorado, Arkansas, so he settled wife #2, Frania Tye Franklin, in Shreveport, Louisiana. He plunged back into speculating in oil leases in Arkansas and Louisiana and did well.

Hunt's Luck and the Daisy Bradford #3

In one of the most famous photographs in petroleum history, a drilling crew poses in self-congratulation at an oil well remote in the Piney Woods of East Texas. In front is the promoter, little Columbus Marion "Dad" Joiner, solemnly shaking hands with his big smiling geologist, A. D. Lloyd. Behind them is the Daisy Bradford #3 oil well, with a wooden derrick and machinery that belonged in a junk pile and run by burning old tires because they couldn't afford fuel. The driller Ed Laster stands off to one side in a white shirt, and the mud-spattered, oil-stained drilling crew crowds around. After three years of failure and frustration, of being paid with promises and worthless shares of stock, of being laughed at for their poor-boy drilling machinery, on this day they have finally brought in an oil well.

The only one without a good reason to be in the picture was H. L. Hunt, impeccably dressed as always, his mouth around a big cigar (hopefully unlit), and a straw boater hat tilted to one side. It was said of Hunt that whenever oil was discovered, he would be there leasing land nearby before anyone else knew what happened. But what stroke of luck drew Hunt to this seemingly unremarkable oil well in the middle of nowhere?

Hunt couldn't know it—no one could—but this well spurned by all the big companies was about to unleash a frenzy of drilling, a flood of oil from thousands of wells. Columbus Joiner would be to wildcatters what Christopher Columbus was to explorers. Joiner had discovered the oil-field equivalent of a vast new continent. His Daisy Bradford #3 was the discovery well for the East Texas oil field, the largest in the lower forty-eight states.

No Hollywood screenwriter would dare make up such an outrageous trio as wildcatter Columbus Joiner, his geologist A. D. "Doc" Lloyd, and lease hound H. L. Hunt. Joiner was a wizened old wildcatter who had failed repeatedly. His greatest talent was sweet-talking old ladies in Dallas into investing in his wells. He told a group of wildcatters: "Every woman has a certain place on her neck, and when I touch it, they automatically start writing me a check."[67] Hunt and Joiner became instant friends. They both dressed well, and both were charismatic and particularly attractive to women. Dr. A. D. Lloyd had been born Joseph Durham but had changed his name repeatedly to elude the six women he had married and abandoned, along with their children. He had numerous careers, including traveling snake-oil pitchman, before he settled on his role as Dr. A. D. Lloyd, geologist. He was no doctor, and his geological training was just enough to master the jargon of lies he made up to sell shares in wildcat wells. And nowhere were his lies more outrageous than the ones he made up about the Daisy Bradford property.

Hunt was in exactly the right place at exactly the right time: the Daisy Bradford #3 on the day the well was tested. When the crew pulled the testing tools out of the hole, oil and mud briefly shot up over the top of the wooden derrick, but the well was no gusher. It settled in at a respectable but unsensational 300 barrels per day. Hunt's luck drew him to the well, then his business skill took over.

Dad Joiner had found oil in the middle of nowhere, but nowhere had advantages. The big companies could not believe that a pair of scammers like Joiner and Lloyd, with their crazy schemes, shady practices, and drilling equipment one step up from junk, could have found anything worthwhile. But H. L. Hunt realized that there was a huge expanse of undrilled land around the Daisy Bradford well—land that might also have oil. For years Joiner kept the project going by selling shares in the well and had sold the well many times over. He was in big legal trouble and badly needed money.

Just as in poker, Hunt was quicker than others to see the opportunity, and he was almost ready to shove all his chips into the pot. While Hunt negotiated on and off with Joiner in Dallas, his scouts watched a well being drilled by Deep Rock Oil a mile west of the Joiner discovery.

About 8:30 p.m. November 26, Hunt's scout phoned in from the field, and dealt him the final card: the Deep Rock well had cut through oil-saturated sand. Hunt acted immediately, and shortly after midnight on November 27, 1930, he and Columbus Joiner signed an agreement by which Hunt bought much of Joiner's leasehold. The price was $1,335,000, but Hunt paid only $30,000 down, money he had borrowed from his haberdasher—the rest would come from oil production.[68] The agreement was one of the greatest coups in the history of the oil business, as Hunt's new leasehold included a fair chunk of the East Texas field.

Hunt made out better than either Joiner or Lloyd. Dad Joiner got a Mexican divorce from his wife of fifty-two years, and at age seventy-three he married his twenty-five-year-old secretary. She stood by him as he drilled dry hole after dry hole trying to find another East Texas field, and was with him when he died broke in 1947. When newspapers printed a picture of Doc Lloyd, his ex-wives and children came running to East Texas to get some money out of him before he disappeared again. He survived his exes but went to prison for fraud related to another oil-well promotion.

After East Texas

The East Texas oil field made H. L Hunt wealthy, but he didn't stop amassing money. For him, it was only the start.

By 1930, most oil companies had staffs of geologists and geophysicists, but H. L. Hunt was slow to use geology. He relied on the lay of the land, called "creekology," looking for patterns in the valleys or creek bends that resembled those of successful oil wells. He was convinced that his success was due to clairvoyance. When his son William Herbert Hunt graduated with a degree in geology in 1950, he became the first geologist working full time for Hunt. Hunt did not hire a geophysicist until 1953.[69]

Hunt believed in luck. When driving with his family, he would stop alongside grassy fields to have everyone search for four-leaf clovers; he himself had a talent for finding them. This may seem like just a game, but his daughter Margaret wrote that H. L. had a "strong superstitious belief" in four-leaf clovers.[70]

Hunt did not amass his great fortune just on luck and four-leaf clovers, although he gave these things some credit. H. L. Hunt was a canny businessman who applied his calculating and poker skills to build a business empire. He was convinced that his success was due to a combination of superior intelligence and extrasensory perception.

According to writer Ruth Knowles, a drilling contractor named Jimmy Owens called on Hunt and told him that he had a dream about drilling a great producing oil well for Hunt in a certain Louisiana location. Hunt immediately agreed not only to drill the well but to give Owens a 50 percent carried working interest—an extraordinarily generous arrangement. The well hit oil and was the discovery well for a fifteen-million-barrel field.[71] This appears to be the Maxie field in Acadia Parish, Louisiana.

Clairvoyant Hassie Hunt

"I've got extrasensory perception [ESP] myself but Hassie had it stronger than anybody."

—H. L. Hunt[72]

Hunt's eldest son, H. L. Hunt III, known as "Hassie," grew into his father's look-alike; he also had his father's lightning mind. H. L. saw Hassie as the one to continue the family dynasty. He had little time for his other children, but H. L. would drive young Hassie around the oil fields of East Texas and encourage his son to recommend spots to drill; Hassie was right enough times to convince Hunt that his son could find oil through ESP just by looking at a map. Hunt once admonished his geophysical staff: "My son Hassie can find more oil with a road map than you so-called educated fellows can find with millions of dollars' worth of equipment!"[73]

When Hassie was nineteen, he started buying and selling oil leases on his own. Like his father, Hassie became expert in buying leases in the paths of expanding oil fields. His sister remembered that charismatic Hassie had many friends who would phone him with news of oil discoveries. When Union Oil discovered the Tinsley oil field in Mississippi in 1939, before Hassie turned twenty-two, Hassie Hunt had oil leases next to the discovery, and he eventually owned thirty wells in the field.

With such a bright and clairvoyant oil-finding son, there seemed to be no limit to the Hunt fortune. But Hassie became increasingly irrational and destructive. He was discharged from the Army during World War II on mental grounds. He was highly intelligent and on a good day could be a very effective manager, but his bad days got worse and more frequent. In desperation, Hassie Hunt underwent a lobotomy in the late 1940s. The operation turned him into a ghostly figure wandering the Hunt estate, spending most of his time sitting quietly in a lawn chair. Even lobotomized, Hassie had occasional flashes of clairvoyance—or so believed his father. H. L. Hunt spent the rest of his life trying to bring the fire back into Hassie's personality.

Consulting Psychics

Because Hunt believed that his ESP was partly responsible for his success, he looked for people gifted to an even greater degree. In the 1930s he would visit Arbie R. "Mighty Red" Chadwick, an African American clairvoyant in Longview, Texas. The seer would not divulge the advice he gave Hunt. Chadwick claimed that during his psychic career he had found fifteen oil fields but would not say for whom.

Hunt met Washington, DC, psychic Jeane Dixon at the Kentucky Derby in 1967 and was impressed when she correctly predicted that Proud Clarion would race from post position seven and would win the Derby. According to Dixon's account, Hunt phoned her the following year for another Derby prediction, but Dixon could predict only "a mix-up of some sort." That year, Dancer's Image won the race but was later disqualified for having been given a painkilling drug. Dixon began advising Hunt on oil leases.[74] Hunt had found his prophetess. He wrote: "As a seeress she's fantastic—and she's without equal. To say that her predictions are amazingly accurate, in my opinion, would be the understatement of the year."[75]

According to Hunt, his geologists recommended that he bid on four tracts offered in a federal lease sale in offshore Louisiana. A month after Dixon impressed Hunt by naming the Derby winner, he asked her which of the four tracts would be the best, and she told him to be sure to win the 5,000-acre Block 207. This was Ship Shoal Block 207, for which a group of seven oil companies headed by Hunt Oil and Placid Oil (both Hunt-owned companies) were preparing a sealed bid for the June 1967 lease sale. Hunt walked into a meeting called to set the bids, pointed to Ship Shoal 207 on a map, and without explanation announced, "I want that one." The Hunt group's sealed bid of $32.5 million set the record high bid at the time for a federal offshore lease and was nearly twice the next highest bid on the tract. One geologist described the tract as "a rather insignificant shallow low-relief anomaly that had been ignored by several of the majors." But there must have been a great deal of geological promise, for the five non-Hunt companies in the consortium also agreed to share the high bid. Both the geologists and Jeane Dixon were correct. Ship Shoal 207 contained a 100-million-barrel oil field.[76]

Hunt also wrote that Dixon recommended that he stay out of a joint venture, saying that it would end in problems and disputes. Hunt stayed out and reported that Dixon had been spot-on.

Hunt used Dixon to try to tap into the telepathic oil-finding ability of his lobotomized son Hassie. Dixon reassured Hunt that Hassie would recover from his lobotomy, but Hassie died in 2005 without returning to a functional personality.

One of Hunt's sons credited Dixon with warning H. L. Hunt away from buying offshore oil leases in the Santa Barbara Channel in the 1960s. A well on one of the leases blew out in 1969, causing one of the most infamous offshore oil spills in the United States. According to another source, however, Hunt declined to bid on the California leases because he remembered the destruction of the San Francisco earthquake of 1906.[77] At any rate, many oil wells have since been safely and successfully drilled in the Santa Barbara Channel, and the one blowout could have been prevented by setting deeper casing.

It's easy to laugh at Hunt's eccentricities, including his trust in hunches, dreams, psychics, and four-leaf clovers, but it worked marvelously for him. It is impossible to say how much of his success was due to his business skill, and how much to his belief in psychics and intuition. Hunt gave credit to both.

H. L. Hunt died in 1974, and leadership of the Hunt companies passed to three of his sons: Nelson Bunker Hunt, William Herbert Hunt, and Ray Lee Hunt. They led more conventional lives and looked for oil in more conventional ways. If his sons drilled wells based on dreams or psychics, they kept quiet about it.

ZULAH LARKIN'S VISION OF OIL

In 1957, oil trade periodicals announced that a new oil field had been discovered in Southern Michigan, at what the magazines drily noted was a "non-geologic location."[78] Was it ever! Choosing where to drill was about the most non-geologic thing possible: the well was drilled, at great hardship to the owners, and against all good advice, because of a vision by a psychic, Zulah Larkin.

The discovery by a previously unknown psychic from Coldwater, Michigan, turned into the Albion-Scipio trend, which has produced more than 150 million barrels of oil. It wasn't found by any geologist, because there was no technology in 1957 that could have found the field, and geologists didn't even know to look for an Albion-Scipio-type oil field.

Back in October 1950, Zulah Larkin woke at 3 a.m. and saw a vision of three people standing in a field, their hands dark with goo. She knew two of the people: Ferne Houseknecht and her brother George Houseknecht. The third person she didn't know, but he wore a striped billed cap. When Larkin wrote to Ferne Houseknecht in Detroit about her vision, Ferne recognized Larkin's description as a place on a seventy-four-acre farm in Hillsdale County, Michigan, which she had bought at auction three months earlier by mistake. She was trying only to bid up the price for her sister and brother-in-law, who were selling. Ferne Houseknecht agreed that the black sticky stuff of the vision must be crude oil, although in 1950, no oil had ever been found in the county.[79]

Zulah Larkin's vision had come effortlessly, in a flash. Then came the long and arduous part: finding a driller, raising money, and drilling the well. It took years and would never have been accomplished without the Houseknechts' faith in Larkin, who had no national reputation or even local fame. She put small classified ads in the Battle Creek *Enquirer* to notify the public that she was a "licensed medium" available for consultation by appointment.[80]

Larkin had been a friend and trusted psychic advisor to the Houseknecht family for many years, first to George and Annette Houseknecht, then after George died to Annette and her grown children: young George (owner of a woodworking shop), Ferne (a sales executive with the Yale Lock Company), Blanche (florist), Helen (housewife), and Luluah (Army nurse). The Houseknechts took Larkin seriously when she predicted that the oil field was worth $547 million, but they were people of modest means with no knowledge of oil drilling. Larkin said that the third person in her vision, the man with the striped billed cap, would drill the well.[81]

Four years later, the Houseknechts drove out to where a drill rig was boring a water well, and there, running the rig, was its owner Clifford Perry, wearing a striped billed cap, as seen in Zulah Larkin's vision. To the Houseknechts, this proved that the vision was correct, that the oil was there, and that the well must be drilled. The Houseknecht children and their mother pooled what money they had and borrowed more from friends and relatives. Ferne Houseknecht persuaded some of her Yale Lock coworkers to invest.[82]

The drillers started boring the Perry Houseknecht No. 1 in May 1955. There was not a single oil or gas well in Hillsdale County, and geologists saw no reason to drill this one. Scoffers called it the Crystal Ball Oil Company. When the well reached a depth of 1,300 feet without finding oil, the investors asked Mrs. Larkin how deep they should drill. Nearly a mile, she answered. The Houseknechts raised more money.[83]

The well sputtered down in fits and starts. Clifford Perry and his crew would drill a hundred feet or so until the money ran out, then move the drilling rig off and drill other jobs until the Houseknechts came up with funds to drill another hundred feet. George Houseknecht mortgaged his plant machinery. Ferne lost her job with Yale Lock and could not contribute much. Investors outside the family dropped out. Ferne Houseknecht later recalled "It was awful. . . . Everyone laughing at us." A well that should have taken less than a month to drill dragged on for two years, but the Houseknecht family stuck together, and still believed in Zulah Larkin's vision.[84]

In September 1956, the borehole hit the top of the Trenton Limestone at 3,650 feet and blew gas for four hours before Clifford Perry could stop the flow. They had a discovery on their hands, perhaps big, perhaps too small to be commercial, but Perry's cable-tool drill rig had reached its limit. The Houseknechts found the money and hired a larger and more expensive rotary drill rig to finish the well. On January 7, 1957, at 3,776 feet and drilling through the Black River Limestone, oil shot up the well and sprayed 500 feet around the drill rig; the well was capped until oil storage tanks could be installed.[85] The drillers stopped at 3,900 feet in June 1957, and the well began producing 140 barrels per day.

The oil flow at the Houseknecht discovery, named the Scipio oil field, was not large enough to create a sensation, yet it encouraged others to drill nearby.

The fourth well in the field, the Stevens No. 1, came in at 2,410 barrels per day in September 1957, and jolted the American oil industry into action. The boom was on.[86]

Oil companies sent agents to Southern Michigan to buy oil leases on all the land for miles around and moved in drilling rigs and crews. The shape and size of the new field were unknown. None of the scientific methods known at that time worked at the Scipio oil field. The limits of the field could only be discovered by systematically offsetting one oil well by another, along with a certain amount of risky wildcatting, until the field was defined by a ring of dry holes.[87]

By June 1958, Scipio field had nine producing oil wells that proved that the field stretched at least a mile and a half long. Seven dry holes showed that the field was only half a mile wide, but how long the field stretched both to the northwest and the southeast was anybody's guess.[88] Drillers eventually bore a continuous belt of oil wells that proved that it was all one long, straight, thin oil field, the Albion-Scipio trend, twenty-nine miles long and one-half mile wide, which produced more than 150 million barrels of oil and 225 billion cubic feet of natural gas.

The Houseknechts profited only modestly. Hundreds of oil wells were drilled along the Albion-Scipio trend, but the Houseknechts owned only Ferne's seventy-four-acre farm. Although they thought that they had handshake agreements with neighboring landowners, their neighbors leased to other oil companies. Under Michigan's regulation of one well for every twenty acres, the Houseknechts could drill four wells on their land.

George Houseknecht had helped reduce expenses during the drilling by working as a roughneck on Clifford Perry's drilling crew. After the discovery, he formed Houseknecht Oil. He drilled one dry hole in 1958, then from 1961 through 1971 he drilled forty-one wells: twelve wildcats (all dry holes) and twenty-nine development wells in Scipio and North Adams oil fields, eighteen of which were completed as oil wells and eleven were dry holes. It is not known if he chose any drilling locations with help from Zulah Larkin, but his record of success does not seem out of the ordinary.

Zulah Larkin died in 1967. Her crazy vision found the largest oil field in Michigan. But, as far as is known, she never found another drop of oil; as far as is known, she never tried. It is easy to dismiss her one-time success as a lucky coincidence, but it was one hell of a coincidence.

BEWITCHING LAURIE CABOT

Don Washburn was convinced that oil lay under a spot on the Grabau farm in Linn County, Iowa. He kept his oil-finding method secret, but it was said to be a machine invented by his friends in California.[89] Washburn admitted that the way he found oil was "goofy and totally unorthodox," yet claimed that he had tried the method on 135 oil wells and had been correct 100 percent of the time.

Washburn told a reporter, "I either have no brains or a lot of guts."[90] He leased oil rights from the Grabau family, hired a driller to drill a thousand-foot hole, and progress on the Grabau #1 well began in December 1978. To supplement his oil-finding machine, Washburn consulted five psychics. One of them was the witch Laurie Cabot.

To find an oil witch, why not start with a real witch? Laurie Cabot is a self-professed witch, and the Official Witch of Salem, Massachusetts, a title bestowed to her in 1977 by Massachusetts Gov. Michael Dukakis as part of a Paul Revere Award for meritorious service. Being the official state witch didn't pay the bills, so Cabot lived and raised her two daughters by running a witchcraft shop in Salem and teaching a course in witchcraft at a local college.

From her home in Salem, Massachusetts, Cabot used her witchcraft to confirm that Don Washburn would find oil. She saw a "river of oil" flowing from Central City, Iowa, to Dubuque, forty miles to the northeast.[91] The well had not found oil when drilling stopped at 1,095 feet in June 1979.

A couple of months later, however, Cabot was back in the hunt for oil in Iowa. She told reporters that she would turn her psychic power to finding oil. An unnamed major oil company had asked her to give them four places to drill in Iowa, agreeing to pay her $200,000 if successful oil wells were drilled on her locations, plus a share of the oil production. Cabot was sure that the company would find oil, explaining: "Yeah, of course they will. Magic works."[92]

We don't know if any of Laurie Cabot's oil-well sites were ever drilled. No oil wells were started in Iowa from 1979 through 1981. Then oil companies drilled seven test wells from 1982 to 1984, all dry holes. If any of those locations were witched by Laurie Cabot, no one ever announced it.

RICHARD IRELAND: PSYCHIC TO THE STARS

Richard Ireland was a spiritualist minister, "psychic to the stars," and psychic oil finder. He was the pastor of the University of Life Church in Phoenix, Arizona, which he founded in 1960. He supplemented his pastoral income with his mentalist act of blindfold billet reading, in which he would blindfold himself and answer written questions in sealed envelopes. He performed in Phoenix nightclubs, occasionally in Las Vegas, and on television talk shows.

Ireland made numerous self-aggrandizing claims. He claimed to have worked for the CIA. He said that his predictions were 92 percent accurate. He listed many academic institutions that had scientifically studied him, including "the University of Vienna; the University of Panama; the University of New York; Washington State College; the Parapsychology Foundation of San Diego." But a search found no controlled studies of Richard Ireland's psychic powers. Neither is there any confirmation that he worked for the CIA. He claimed to hold four doctorate degrees, including one from Arizona State University. A reporter found a number of Ireland's claims to be false, including his degree from Arizona State.[93]

He became a "psychic to the stars." His most notable client was Mae West, by then a fading caricature of a sexpot, who gushed that Ireland was "the greatest." The admiration was mutual, and after she died, Ireland told a reporter: "I don't want to tell you that I have been in her bed before, but I have been in her bed before."[94] He also advised Hollywood television actors Amanda Blake (*Gunsmoke*) and Eddie Albert (*Green Acres*).

Richard Ireland used his psychic powers for Dallas oil man Jerry Conser. In August 1980, Conser hit oil at one of the locations Ireland had approved in Nolan County, Texas, and immodestly named it the Ireland-Conser oil field. A video filmed in 1981 has Conser's chief geologist telling the camera that the Ireland-Conser field discovery well started out flowing more than a thousand barrels of oil per day, and Conser said in 1981 that the field held "certified reserves" of a hundred million dollars in oil.

The value of the Ireland-Conser oil field was outrageously exaggerated. It was a one-well field, and the completion report filed with the State of Texas shows that the one producing well, the #1 Moore, began at 206 barrels per day. The well was plugged after producing 194,000 barrels of oil and thirty-nine million cubic feet of gas, an output worth less than five million dollars. Conser tried to extend the field, but stopped after drilling four dry holes.[95]

Between the discovery of the Ireland-Conser field in August 1980 and Richard Ireland's death in January 1992, Conser's companies drilled another fourteen holes. Only one of the wells appears to have been profitable. The rest were either dry holes or produced less than 900 barrels of oil: clearly not enough to pay for the well.

A YouTube video shows Ireland being interviewed at the Millennium #2 well, while it was drilling in Mills County, Texas, in 1981. He predicts on camera that the well will find oil, and the Millennium Minerals chief geologist enthuses that it is a good prospect because granite is under the location at a shallow depth.[96] Most geologists would say that it was a very poor place to drill for that very reason. It was a dry hole, another failure for Richard Ireland.

DOC ANDERSON AND IVAN THE TERRIBLE LOOK FOR OIL

On a windy November day in West Texas, psychic "Doc" Anderson stepped out of the car and walked across the land with eyes closed, arms extended forward. A television camera crew from Chattanooga followed him. Anderson, a big bear of a man with a goatee beard, stopped walking and started shaking; he turned his palms upward—they were bleeding, blood dripping to the ground. This was the place to drill for oil. Robert C. "Doc" Anderson, psychic counselor to television stars, was far from Hollywood, but his oil-finding was more show business than oil business.

Anderson's partner, John R. Shaw, had needed investors for his oil drilling. The best way to attract investors is to drill money-making oil wells, but Shaw drilled mostly dry holes, so he needed a gimmick. According to a melodramatic passage in a 1973 biography of Anderson, in November 1968, John Shaw was so despondent after a string of dry holes that he tried to shoot himself in the head in a Los Angeles hotel room and was saved only because his .45 pistol misfired. Later that day in the Los Angeles airport, a stranger handed him a magazine with an article on Doc Anderson. Shaw read the article and switched his ticket to fly to Chattanooga to see the famous psychic.

Before entering the oil business, Shaw had been a Dallas policeman before becoming a professional wrestler. As a policeman, he used to hang out at a nightclub owned by Jack Ruby, who always welcomed policemen. When Lee Oswald killed President John F. Kennedy in Dallas in 1963, Shaw was in Midland, Texas, trying to sell shares in some wells he wanted to drill. Shaw's investors backed out of the deal after the president was shot, sending John Shaw on a liquor binge. Several days later, after Jack Ruby had in turn killed Oswald, Shaw (still drunk) sent a telegram congratulating his old acquaintance, Jack Ruby. Shaw's impulsive bad judgement earned him a grilling by an FBI agent.[97]

Shaw wrestled as "Ivan the Terrible." When he became an oil man, scowling beneath a thick beard and a fur cap would not have attracted investors. He needed a new gimmick, and his show-business mind struck on the idea of using a psychic to attract investors.

Like John R. Shaw, Robert C. Anderson had a long career in show business, although in large part, we have only Anderson's version of his early life, which he appears to have embellished more than a bit. Born in Iowa in 1908, he was a coal miner, prizefighter, fisherman, circus strongman, and professional wrestler. By his own account, he travelled the world from Tibet to Africa, and became the "undisputed world's champion" in "Roman-style" bullfighting.[98] There is no corroboration that Anderson was ever in a bull ring, or even that there is such a thing as "Roman-style" bullfighting, but Anderson handed out postcards of himself in bullfighting costume.

Anderson married, had children, quit the circus, and became the travelling psychic "Doc" Anderson—although he advertised himself as an astrologer, presumably to avoid local ordinances against fortune-tellers. He would stay at each town for a week or two, placing an ad in the local paper that the "world famed astrologer" was staying at such-and-such hotel and available for consultation. He made bold but scattershot predictions, knowing that his errors would be forgotten, and if one of his predictions hit, he could brag about it. In August 1942, he told a newspaper editor in Murray, Utah, that Germany would sue for peace with Russia by October 1942, then by February 1943, Germany would surrender to the Allies and declare war on Italy. In October 1942, Anderson predicted that Germany would be defeated in December 1942.[99] Anderson later claimed to have accurately predicted the death of Franklin Roosevelt and the dates of the end of World War II in both Europe and the Pacific—and given his habit of making multiple predictions, he may have gotten a few right.

In 1944, Doc Anderson settled in Rossville, Georgia, in a house one block south of the state line and Chattanooga, Tennessee. His office was also in Rossville, right up against the Tennessee state line and Chattanooga city limits.[100] Living and working in a different state from most customers offered a measure of legal safety in the suspect profession of fortune-telling.

Doc Anderson gave his time and money to community causes. In 1971, he gave $6,000 to help a local woman get a life-saving kidney operation. Several years later the Rossville Lions Club gave their first annual Citizenship Award to Doc Anderson.[101]

Anderson gained national press in the 1960s and made one failed prediction after another. In 1967, he proclaimed that George Wallace would be elected president in 1968—a prediction that he repeated for the 1976 election (Anderson insisted that he had predicted correctly, but that Wallace's victory had been stolen). He predicted in 1967 and again in 1977 that, within two years, the US would be at war with China. Anderson predicted life on the moon and described natives of the planet Venus as meek people with greenish skin and slanted heads. He said that by 1995 a Convention of United Planets would take place in Virginia, where earthlings and space aliens would draft a treaty for space travel.[102] His record as a seer was miserable, but his dramatic predictions earned him publicity in year-end surveys of psychic predictions.

Anderson made good copy on slow news days, such as when he was photographed shattering a wineglass on the table in front of him just by staring at it. Local newspaper and TV reporters built up their hometown psychic star and always mentioned his few successful predictions from the 1940s but neglected his many well-documented recent failures. Chattanooga newspapers printed pictures of Doc Anderson with Hollywood celebrities: Doris Day, Burl Ives, Denver Pyle, and Eddie Albert. Actor Clint Walker stayed in Anderson's home while playing in a celebrity golf tournament.

In 1968, John R. Shaw started tagging along with Anderson on visits to Hollywood looking for oil investment money. Anderson had entrée as a psychic advisor to Hollywood stars and promised to use his psychic power to find oil for Shaw's drilling. Shaw was one of many oil men looking for investment money in Hollywood. He let people assume that he was wealthier than he was—no one wants to invest money with a poor oil man. He developed a crush on Doris Day, and got drunk, tripped, and injured himself while escorting Day's mother to a party. Day charitably visited him in the hospital the next day and was angered to discover his wife there with him, for Shaw had hidden the fact that he was married. Doris Day soured on John Shaw, but other actors invested in his oil drilling ventures, including George Raft.[103] With Doc Anderson picking places to drill, how could they miss? Shaw told investors that Anderson's oil locations had produced an unbroken string of successful oil wells.

By the time he joined Shaw to look for oil, Anderson was an overweight sixty-year-old, with a moustache, sometimes a goatee beard, a gentle manner, and an engaging smile. A television news film from 1968 shows Anderson prospecting for oil by walking across a field in West Texas. His eyes are half shut and his jaw slack; his body shakes; his forearms are extended forward at the elbow, palms-up. His palms are bleeding. It looks like a scene from a low-budget horror film. Bleeding palms, called *stigmata*, in imitation of Christ's wounds, have been experienced by Christian mystics for centuries, but Anderson was probably the only one to use stigmata to find petroleum.

In 1972, oilman John Shaw and television actor Denver Pyle told the Chattanooga newspapers that they wanted to erect a life-size statue of Doc Anderson in some public place in Rossville. They would pay for the statue; all they wanted was city permission and a place to put it. The mayor said that the commission would consider the proposal, and that was the last anyone heard of it.

Trouble with the law came in 1976, when the US Securities and Exchange Commission (SEC) filed a civil suit to stop John Shaw from selling unregistered securities. The commission said that Shaw had raised more than three million dollars from investors since 1971 but had diverted much of it to his personal use. The suit also said that although Shaw bragged to investors that Doc Anderson had been "100 percent successful" in locating oil wells, none of the wells had been profitable and all his investors lost money.[104] Shaw settled with the SEC by agreeing not to sell unregistered securities.

In 1978, the US Postal service charged John Shaw and Doc Anderson with mail fraud. Postal authorities said that Anderson and Shaw had defrauded some 300 investors by promising that Doc Anderson was infallible in finding oil. Anderson and Shaw liked to show pictures of themselves with Hollywood stars such as Eddie Albert and Doris Day, but when questioned, Day and Albert confirmed that they knew Anderson but denied investing in his oil locations.

Shaw and Anderson pled guilty. Shaw served six months of a three-year sentence. Anderson, because of age and ill health, got off with probation.[105] The Chattanooga newspapers printed nothing about Anderson's legal troubles.

In December 1979, at age seventy-two, Doc Anderson announced that he was training to face the fighting bulls in Mexico City the next March. He never got to Mexico City. On the morning of March 21, 1980, Doc Anderson drove his Lincoln Continental toward his office in heavy rain. A few blocks from home, he tried to drive through Chattanooga Creek where it was flooding across the road. His car stalled, and when he tried to wade to high ground, the current swept him away.[106]

Doc Anderson was much beloved around Chattanooga. Both daily newspapers reported his apparent drowning with big front-page headlines and followed the search by scuba divers until the body was discovered several days later.[107] His funeral overflowed the church. Conspicuously absent were his former Hollywood friends, but his partner-in-crime John Shaw came from Texas to pay his respects.

John Robert "Ivan the Terrible" Shaw, ex-policeman, ex-wrestler, and ex-con, died in Dallas in 1989.[108]

W. W. KEELER: AN OIL EXECUTIVE HIRES PSYCHICS

One of the most prominent oil men to use psychic methods was W. W. Keeler, the chairman of Phillips Petroleum, then the ninth largest oil company in the US in both refining capacity and gasoline sales.[109] Keeler was one of the few major oil executives in the late twentieth century who openly believed in psychics.

William Wayne Keeler was born in 1908. He worked summers for Phillips Petroleum until he graduated from the University of Kansas in 1929, when he became a full-time chemical engineer for Phillips. Keeler rose through the ranks at oil refineries and petrochemical plants. He was also an enrolled member of the Cherokee Nation, and in 1949 President Truman named him principal chief of the tribe.[110]

Keeler relied on hunches—he said that it was literally a gut feeling. He recalled when his gut feeling led him to correctly choose the best process for artificial rubber, and another time his hunch allowed him to accurately extrapolate from a single data point the performance of a new catalytic process. Acting on hunches would have been familiar to company founder Frank Phillips, who began as a wildcatter before geology was popular and had used the wildcatters' grab-bag of oil-finding tools, including divining rods, doodlebugs, and creekology.[111]

Keeler became president and CEO of Phillips in 1967, Chairman of the Board in 1968, and began using hunches to find oil. In 1969 he befriended psychic investigator Harold Sherman, who introduced Keeler to a number of psychics. Keeler was convinced that he had ESP to a small degree and wanted to find a clairvoyant with greater powers.

B. Anne Gehman

In 1972, Harold Sherman introduced Keeler to Beatrice Anne Gehman, a young spirit medium and regular at the spiritualist community of Cassadaga, Florida. Gehman impressed Keeler with a psychic reading, so he asked her if she had ever dowsed for oil. Years previous, she had done some oil-finding for a client by passing her hands over a map and sensing places on the map with oil. Her client later told her that drilling had proved her right.

Keeler was enthusiastic about Gehman's oil-finding and introduced her to Phillips management. Harold Sherman optimistically wrote: "we appear to be close to a million-dollar deal," but the other Phillips executives didn't want to hire a psychic. In September 1972 the *Orlando Sentinel* noted that William W. Keeler had come to Florida to consult with Gehman on oil business.[112]

About every four to six weeks Keeler would send a plane to Orlando to fly her to oil prospect locations in the Caribbean and Central America, where she would use her psychic power to advise on drilling locations. Gehman wrote that Keeler told her that her oil-finding was 90 percent accurate. Keeler also had some contact with famous psychic Jeane Dixon but said that he was more impressed with Anne Gehman.

Keeler sent a private jet to fly Anne Gehman to southwest New Mexico. She stretched out face-down on the ground, arms and legs outstretched, and detected oil at 3,600 feet. Keeler had vials of oil of different quality, and Gehman touched each vial, and sensed which vial held oil similar to the oil below. According to Gehman's biographer, Keeler later phoned Gehman with the news that the drill found oil at just the depth and oil quality she predicted. Gehman wrote that she received a large bonus for her work. It's a great story but has the fatal flaw that no one has ever produced oil from southwest New Mexico.

Keeler retired as corporation president in January 1973. In March 1973, he told a Miami *Herald* reporter, "I wouldn't hesitate to ask her [Gehman] questions relating to Phillips 76."[113] That same month he resigned as chairman.

Ingo Swann

Keeler met psychic Ingo Swann in 1973 and hired him to find oil. Swann was a New York City artist with a reputation as a clairvoyant. He grew up as Ingo Swan (one "n" to his last name) in Telluride, Colorado, back when it was a run-down mining town and before it became a trendy ski mecca. After a stint in the Army, he moved to New York, became an artist, and added the second "n" in Swann for numerological reasons. Swann had been tested at the Stanford Research Institute, in California, where he helped design a clairvoyant technique called Remote Viewing.

At Keeler's request, Ingo Swann tested his psychic powers on potential oil drilling locations in September 1973. He began by visiting twelve locations in Tennessee in the company of three oil dowsers: brothers Willie and Earl Pyle, and Earl's wife Sarah Pyle. Swann would describe the subsurface conditions based on his remote viewing, including presence of oil and gas, and the dowsers would compare Swann's visions with their dowsing, or sometimes with known results from drilled wells. After two days in Tennessee with the Pyles, Swann examined oil prospects in Oklahoma, Louisiana, Mississippi, and Texas.

Keeler left Phillips in March 1973, and without Keeler, Phillips Petroleum was uninterested in using psychics. After Keeler took Ingo Swann to examine some Phillips properties in October 1973 in West Texas, Swann asked Phillips for information on the properties, prompting a letter to Swann from an executive vice president of Phillips cutting off all further communication.[114]

Keeler retired from Phillips, but not from the oil business. He drilled wells as Red Feather Oil and Gas, Keeler Investments, and W. W. Keeler and Sons.[115] From 1973 to 1986, Keeler and Red Feather drilled a total of thirty-seven wells with known outcomes in Arkansas, Kansas, Kentucky, and Tennessee, with the majority (twenty-one) in Tennessee. Out of fifteen development wells, ten (67 percent) were completed as oil and gas wells. Out of twenty-two wildcat wells, seven (32 percent) were completed as oil and gas wells. For comparison, during the same period, 83 percent of the development wells and 53 percent of the exploratory wells drilled in Tennessee were successful. Ingo Swann's involvement with selecting drill locations is unknown, but the oil-finding records of Keeler and Red Feather are not out of the ordinary.

Whatever the results, his oil-finding work was apparently lucrative enough for Swann to buy a four-story house in Manhattan.[116] William W. Keeler died in 1987. Ingo Swann died in 2013.

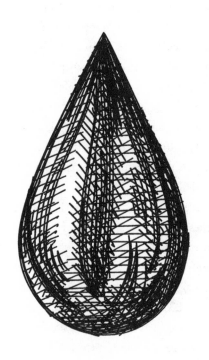

WHAT IS DOWSING?

Dowsing is a search for a hidden substance using either a physical sensation or movement of a simple mechanical device, where the sensation or device movement has no known physical cause in relation to the substance.

The definition includes the search for things underground, including water, metal, oil, and utility pipes. It includes the simple mechanical devices used and excludes science-based exploration methods such as gravity, magnetism, and metal detectors. The dowsing definition also excludes other forms of divination such as astrology, Tarot cards, and dreams. Dowsing began among German-speaking miners in central Europe about AD 1400 as a way to find ore deposits. By the 1600s, it was also being used to find groundwater and buried treasure.

The distinction between dowsing and other forms of divination is blurred by modern diviners who use dowsing instruments to decide any question. The diviner uses his pendulum to predict the future, select books to read, determine food to eat, and communicate with plants, spirits, and energies. These practices are divination, but not dowsing.

For centuries, dowsers needed to tread over the ground in question. Then at a French dowsing congress in 1913, Mr. Mathhieu showed that he could dowse anywhere from any place—all he needed was a map. This astonished the other dowsers, but some discovered that they could map-dowse as well. The innovation spread first in France, then later in Britain, Germany, and the US.

CHAPTER 5

Dowsing Instruments

Original dowsing rods were Y-shaped branches from a bush or tree. The dowser holds the prongs one in each hand, stem pointing forward, until the stem moves up or down. Other flexible materials were substituted for branches, including whalebone, metal, and plastic. For centuries, Y- and V-shaped rods were nearly the only instruments used by dowsers. Y-rods declined in popularity in the late twentieth century, and in the United States, the ability to dowse with a Y-rod is scarce enough to be a sort of status symbol among dowsers.[1]

L-shaped rods are used singly or in pairs, with the short end of the L held upright in the fist, with the long end pointing forward. When the dowser walks over the target, the rods swing. Also called angle rods, they first came into common use by the early 1950s by utility workers to locate buried pipes. They became broadly popular in the US in 1967 when newspapers printed photos of US Marines trained by dowser Louis Matacia to find Viet Cong tunnels by dowsing with bent coat-hanger wire.[2]

Some dowsers dismissed L-rods as inferior instruments. Kenneth Roberts wrote that L-rods were not for the true dowser: "They have something to do with dowsing, but non-dowsers are able to use them." Howard Chambers noted: "The swing rod is really more of a toy than a serious device." Jim Kuebelbeck wrote: "In my opinion, no single instrument has caused more harm or brought more discredit to the credibility of dowsing than the 'L-Rod.'"[3]

Pendulums have been used for divination for thousands of years, and for dowsing since 1641. But the dowsing pendulum did not become popular until the early twentieth century in France, the mid-twentieth century in Europe, and the late twentieth century in the US. The French-born Swiss priest Alexis Mermet did most to popularize the dowsing pendulum. French ethnologist Anne Jaeger-Nosal noted, "I have never seen a dowser-geobiologist without a pendulum in the pocket."[4]

Pendulums are the most popular dowsing instrument in the US. The delay in adopting the pendulum in America, compared to Europe, may be that previous to the start of the American Society of Dowsers in 1961, American dowsers were more likely to be guided by folk traditions.

A bobber is a single stick-like instrument, which is either flexible and held at one end, or rigid and held balanced in one hand.

Some dowsers rely on physical sensations rather than instruments. The sensations, sometimes painful, seem to be unique to each dowser. An oil smeller in Wyoming said that his feet tingled over an oil deposit. In Nebraska, L. F. Heseman reacted to oil by becoming nauseous, and the more oil beneath him, the more nauseous he became. A Florida man detected oil by the way his stomach rumbled. A South Dakota pickle maker said that he could find gold, silver, iron, coal, and oil because each caused pains in different parts of his body. Chet Boogher reacted to gas by a feeling in his throat which ranged from a tickle to a choking sensation.[5]

The popularity of dowsing instruments has changed through the years. From the start of the oil industry in 1859 until the late 1880s, the Y-rod was nearly the only instrument used in oil dowsing. Y-rods are still used by about half of oil dowsers. From the late 1880s to the end of the century, using physical sensation without instruments was common. Oil dowsing with pendulums became very popular about 1900, but in the 1980s and 1990s, pendulums were largely replaced in field dowsing by L-rods. From the 1980s onward, field dowsing appears to be split roughly between Y-rods and L-rods, while map dowsing appears to be mostly done with pendulums. Many oil dowsers have a two-step process, starting by map dowsing with a pendulum, then confirming exact locations by field dowsing with an L-rod or Y-rod.

American Dowsers
The American scientific establishment rejected dowsing. Professor Benjamin Silliman of Yale College concluded in 1826 after some failed dowsing experiments: "The pretentions of diviners are worthless. The art of finding fountains and minerals with a succulent twig, is a cheat upon those who practice it, an offense to reason and to common sense; an art abhorrent to the laws of nature, and deserving universal reprobation."[6]

But dowsing was too deeply rooted to wither in the face of mere academic disapproval. When the first oil boom began in 1859, dowsing was ready to search for a new treasure: petroleum.

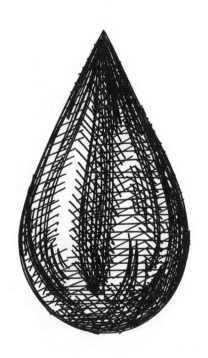

EARLY OIL DOWSING

WHEN EDWIN DRAKE FOUND OIL AT TITUSVILLE, PENNSYLVANIA, water dowsers quickly adapted their craft to oil. Some searched for oil with the same forked twigs that they used for water, but others adapted the equipment for oil. Some dipped the free tip of the rod in oil; sometimes a bottle of oil or an oil-soaked cloth was attached to the tip. Dowsing pendulums were often bottles of oil at the end of a string. Sometimes the sample was not crude oil, but a secret chemical mixture said to have an affinity with crude oil.

> "There are, perhaps, a dozen of these greasy wizards perambulating Venango
> county, and locating wells after this fashion, in consideration of a ten-dollar
> bill, and what is stranger, there are plenty of people with the ten-dollar bill."
> —Titusville (Pennsylvania) *Morning Herald*, August 22, 1865

Oil dowsers were common in the 1860s and 1870s, but newspapers usually left them unnamed. One of the named oil dowsers was Thomas H. Brown, who in 1864 dowsed the location for the discovery well for the Pithole oil field. But Brown's hazel twig didn't flinch for many nearby properties later found out to be rich in oil. Another well-known dowser was Captain P. Smith who dowsed successfully at Pithole, Tidioute, and Petroleum Centre.[1]

Even many skeptical drillers hired dowsers because they didn't charge much, and if they could find oil, so much the better. Out of some 10,000 to 12,000 wells drilled in Pennsylvania up to 1871, at least 1,000 were

Fig. 6. Satirical take on an early oil dowser, also known as an oil witch. (Samuel Gamble Bayne, *Derricks of Destiny* (New York: Brentano's, 1924), 53.)

said to have been guided by various psychic methods. The diviners and oil smellers typically charged $10 to $100 and were said to have hit about as often as they missed. Frederick Crocker agreed to pay a dowser the first three days of production: his Rattlesnake well came in strong in 1867, and the dowser walked away with $1,500. William Wright noted in 1865 that

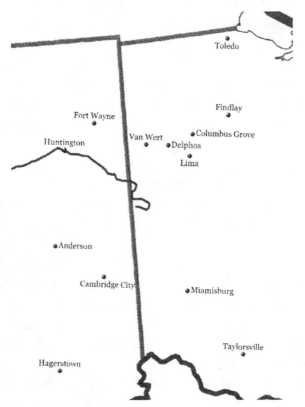

Fig. 7. John Booher's well locations in the Ohio–Indiana
gas trend.

some of the best wells had been located by oil smellers, but that in most
cases, the prophecies failed.[2]

The situation was similar in the new oil fields of the Ohio Valley, in Ohio,
Kentucky, and what soon became West Virginia. Oil dowsers were particularly
common in the Macksburg field (discovered 1861) of Ohio, although most well
locations trusted in pure luck.[3]

Some early oil hunters were known as "oil smellers," similar to water dowsers
being known as "water smellers," a term still used in Pennsylvania. But "oil
smeller" was broadly used. Geologists were commonly called oil smellers into
the twentieth century. Some oil smellers said that they could actually smell the
oil, and in the early years of drilling shallow wells above oil seeps, a keen sense
of smell could be rewarding. One smeller sued a golf companion whose golf ball
had hit and broken his nose, depriving the smeller of his livelihood.[4]

If humans could be "oil smellers" then animals such as dogs, with their more acute sense of smell, should be so much better. Dogs have been successfully used in mineral exploration to sniff out sulfide mineralization, and it seems possible that a dog's acute sense of smell may detect underground oil and gas. A dog at Van Wert, Ohio, selected locations for successful gas wells in 1888.[5] Some uses of animals were more frivolous. In 1915, an oil man in Texas selected a drilling location by turning his mule loose in a field and drilling where the animal stopped.[6]

BOOHER: THE GAS CRANK

John S. Booher had a short but spectacular career as a human natural gas detector. He was a sensation in the Midwest, but his career only lasted about seven months.

Gas wells drilled near Findlay, Ohio, in 1886 flowed up to fifteen million cubic feet per day, and sparked a gas boom. The drilling frenzy extended the new gas region by random hit-or-miss drilling northeast to Lake Erie, and southwest through Lima, Ohio, and into eastern Indiana. Cheap natural gas attracted industry to Findlay, and every town in western Ohio and eastern Indiana wanted to be like Findlay.

John S. Booher owned and ran a stone quarry at Taylorsville, Ohio. In August 1887 he discovered that whenever he was above natural gas, electricity running up from the ground sent him into painful convulsions. By lifting one foot off the ground, he could break the electric circuit, and return to normal. He went to the #4 well while it was drilling at Miamisburg, Ohio, felt nothing, and correctly predicted that it would be a dry hole. The company had him choose the location for well #5, which came in with a strong flow of gas in December 1887, and Booher won newspaper publicity throughout the gas region.[7]

Booher was the man of the hour to gas-hopeful communities. Booher's friend Major Comer left his cigar store in Van Wert, Ohio, to be his agent and publicist. Comer told reporters that his rank came from being a drum major.

Booher sent letters to towns throughout western Ohio and eastern Indiana, promising to find gas. The mayor of Menomonie, Wisconsin, received a letter from Booher stating that he had felt a gas vein near that town while passing through on a train, and offered his services.[8] If Booher found gas, the company would pay him $1,000 (about the cost of drilling a well); if not, Booher would pay the company half the cost of drilling.

Fig. 8. John Booher searching for gas at Fort Wayne, Indiana, by the electric shocks he felt from underground. (*Chicago Daily Tribune*, February 8, 1888.)

After Booher's well at Delphos, Ohio, tested five million cubic feet per day, the *Chicago Tribune* printed a long, illustrated article about him and his reputation was made. Companies hired him to find gas at Van Wert and Columbus Grove, Ohio, and Cambridge City, Hagerstown, Huntington, and Anderson, Indiana.[9]

Booher was mostly successful as long as he stayed in gas-rich territory, but made a major error when he went to find gas in Minnesota, which was in a frenzy for natural gas. Excavations and water wells had penetrated old bogs where bacterial decay had generated methane gas. The gas occurred only in small pockets that were quickly exhausted, but Minnesota towns were optimistic that they could find long-term supplies.[10]

Businessmen brought Booher to Mankato, Minnesota, where he selected a spot at Minneopa Falls, a few miles from town. Boring started in the summer of 1888, but the drillers struck granite at 504 feet. The drillers pronounced the well a lost cause, but Booher insisted that he could feel the gas, so the drill bit struggled slowly down through granite. The company suspended drilling for the winter when the depth reached 1,000 feet. The directors tried to collect an assessment to drill another 500 feet, but the shareholders had seen enough granite, and in May 1889 the company gave up the attempt.[11]

When Booher went to Minnesota, things were turning sour back in Ohio and Indiana. The gas company at Van Wert, Ohio, was disappointed at the low flow rates of Booher-located wells. By early 1889, they had given up on the wells and planned to pipe in gas from farther away. Booher's location at Anderson, Indiana, proved to be a high-profile dud in May 1888. In June 1888, the gas company at Huntington, Indiana, announced that they would no longer use his services. Booher needed a big success in Minnesota to keep his career flying, but it turned into his most famous failure.[12] After Minnesota, he disappeared from the newspapers.

Booher's early success drew attention to other gas sensitives. E. Alexander, the "Cleveland natural gas wizard," became the subject of news articles in 1888. In Western Pennsylvania, Chet Boogher, another "gas crank," became confused with John Booher.[13]

BERT REESE: PSYCHIC OIL DOWSER FOR THE ROCKEFELLERS?

All over the internet, you find statements that John D. Rockefeller drilled his first oil wells based on dowsing. European dowsing literature names Bert Reese as the one who successfully dowsed for Rockefeller, the founder of Standard Oil.[14] But did he?

Bert Reese was born Berthold Riess in 1840 in the village of Pudewitz, near Poznan, Poland. By his own account, he was mistreated because of his clairvoyance, so he emigrated to America.[15] By 1888, he was touring as Professor Bert Reese, mentalist and psychic advisor, "The Star of Mystery. The King of all prophets!"[16]

Reese's specialty was "pellet reading" or "billet reading," in which people write questions on small slips of paper. Reese would hold each piece of paper against his forehead and answer the question. To some, his demonstrations of blind pellet reading are proof of ESP. But magicians regard Reese as an entertainer like themselves and admire and study Reese's conjuring tricks. Magician Theodore Annemann wrote of Reese: ". . . in the past 30 years one man stood out as a charlatan par excellence at reading the folded clip." Annemann wrote that Reese was a master not only at sleight of hand but also at "sleight of mind," which is planting a suggestion so that people remember things in a slightly different way than they actually happened. Reese would create confusion with his tricks, then suggest to people that things happened in a different manner or sequence than actually took place; in their confusion, most "remembered" the altered version. Creating such false memories seems incredible to most people but is well known to magicians and psychologists.[17]

Reese impressed a number of scientists, including James Hyslop of Columbia University and Harvard psychologist Hugo Muensterberg. Magician Joseph Rinn detected Reese's pellet-reading trick with some difficulty and alerted Harry Houdini to Reese's method.[18] Years later, Houdini, not mentioning any tip-off from Rinn (Houdini was never one to share the spotlight), said that he had caught Reese red-handed in trickery. Houdini complimented Reese (and, of course, himself): "I was amazed at his skill, and if I had not been extremely familiar with all sleights, and all moves of Mediums, who resort to the Pellet Test, I would have been completely fooled."[19]

Reese said that oil operators Vandergrift and Wasserman took him to Western Pennsylvania to find oil, and he located oil wells at Titusville and Petroleum Centre before farmers ran him off at gunpoint for using his psychic powers. A reporter found that Reese's account contained a large dose of fiction, and no one but Reese remembered his oil-finding.

Beginning in the 1920s, European parapsychology literature carried statements that a dowser, sometimes identified as Reese, had found oil for John D. Rockefeller. French paranormal researcher Eugene Osty noted of Reese: "[H]e was a remarkable dowser, to whom Mr. Rockefeller owed some of his most valuable oil deposits."[20] The Austrian Alois Wiesinger wrote: "Professor Bert Reese discovered Rockefeller's petroleum deposits."[21]

Neither man wrote in English: Osty wrote in French, Wiesinger in German, and neither gave the source of information, though it was likely Reese himself. Biographies of John D. Rockefeller do not mention Reese or dowsing, however, and neither do histories of his Standard Oil Company. Perhaps Reese made the claim only in Europe, knowing that he would be contradicted if he made

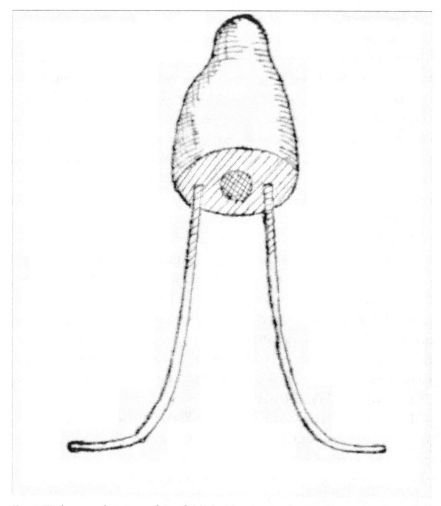

Fig. 9. We have no drawings of Canfield's bobber, but its description was similar to this instrument used by Pennsylvania oil dowser Andrew Thompson. (*Pittsburgh Daily Post*, March 28, 1892.)

the claim in the US. There appears to be no credible source for the story that Bert Reese—or any dowser—found early oil wells for John D. Rockefeller or his Standard Oil Company.[22]

ISAAC CANFIELD: FROM THE CIVIL WAR TO THE JAZZ AGE

Few dowsers have been as successful in finding oil as Isaac Canfield. He accidentally discovered the second oil field in Colorado, then twenty years later he discovered the third oil field in Colorado. He has a claim to the first oil discovery (although non-commercial) in New Mexico, and he dowsed in the location for the Electra field discovery well in Texas. How did he do it? He dowsed the discoveries with an oil-finding rod he called the bobber.

Isaac Canfield inherited the bobber from his father Ira, who got it from an unnamed inventor at an undetermined time. It was a container shaped like half of an egg, with a compartment for the secret oil-attractive mixture of chemicals, and holes for the ends of two wooden rods. The oil dowser grasped the other ends of the rods, one in each hand, and used it the same as a forked twig. The dowser would walk about until the attraction between the chemicals and the underground oil would pull the bobber downward.

Isaac Canfield had been a miller in Coudersport, Pennsylvania, where his father Ira built and ran a grain mill. In 1860, soon after the Drake well hit oil, Isaac and Ira moved to Titusville, Pennsylvania, and entered the new oil industry. In the 1860s, oil made them wealthy. Ira built a showplace house in Titusville and was described as "one of the oil princes."[23] But fortune turned and soon wiped them out. If Ira Canfield used the bobber in Pennsylvania, it didn't do him much good.

In 1871, the two Canfield men migrated west as members of the Greeley colony, the organization that founded the city of Greeley, Colorado. The Canfields saw opportunity in coal mining. Many years later, Isaac recalled that he was excavating in search of coal when he heard a voice say, "Sink twenty-five feet deeper, and you will find a vein of coal seven feet thick." He looked up and saw a man peering down the shaft at him. "Who are you?" he asked. "My name is Hayden. I am making a geological survey of this area." According to Isaac, Hayden's prediction came true.[24]

The Canfields opened the Rob Roy Coal Mine in 1875, northeast of Boulder. They employed more than a hundred miners and started the town of Canfield, Colorado, next to the mine. The Canfields did well until they tried to reduce the miners' wages in 1877. The miners not only refused to work for less than seventy-five cents per ton but decided that they deserved a raise to one dollar per ton. Rather than comply, the Canfields closed the mine and sold it the following year.[25]

Ira and Isaac Canfield moved to Florence, Colorado, a coal-mining town near Cañon City, where they again opened up their own Canfield coal mine.

The Canfields had landed in the one area in Colorado known to have petroleum. Since no one knows when, the Ute Indians knew of an oil spring, where petroleum flowed out of the ground along Oil Creek (now Fourmile Creek), northeast of present-day Cañon City. But no matter what people did, they could not extract more than the few barrels per day that flowed into some hand-dug pits. In 1863, indefatigable oil promoter Alexander Cassiday drilled a four-inch diameter well twenty-three feet into the ground, making it the first oil well in Colorado, although oil production from the well was small. Dry holes were drilled nearby in 1873 and 1877.

In 1878, Isaac Canfield went back into the oil business. Some building excavations in Cañon City had uncovered oil seeps, so Canfield, Alexander Cassiday, Cassiday's son, and some Denver businessmen leased 1,200 acres near the seep. Isaac Canfield drilled down 780 feet on the Macon lease without finding oil. It seems likely that Canfield would have dowsed the location with his bobber, but we don't know for sure.

In 1881, Isaac Canfield joined with Alexander Cassiday to drill the first deep oil well in Colorado, but probably by accident. It was supposed to be a water well for the nearby coal mines. However, some suspected that Canfield and Cassiday were drilling for oil and made up the water-well story to get around oil-averse shareholders. The Grand Canyon well hit oil. It was the discovery well for the Florence oil field, which was the second oil field in Colorado, after Cañon City (the two have since merged into the Florence–Cañon City field).[26]

Canfield and Cassiday incorporated the Arkansas Valley Oil Company in September 1881 and included their partners in the failed Macon well of three years previous. Other local oil companies formed, and hundreds of wells were drilled in Colorado's first oil boom.[27]

Old-fashioned oil dowsers would walk back and forth to find the exact spot to sink a well, lest they miss the "vein" of oil. In most oil fields, the oil is in extensive layers of porous rock, so that moving the location a few feet or a hundred feet or so serves no purpose other than good theater to have the rod suddenly point downward over the supposed vein. But the Florence field is an exception. There, the oil is contained in near-vertical fractures in the Pierre Shale. Hit an open fracture and you get an oil well. Drill a few feet away, you miss the fracture and have a dry hole. Drilling even within the field is hit-or-miss, and in the early days about one well in three within the field was a dry hole. Arkansas Valley Oil did somewhat better, with only one dry hole out of six or seven drilled.[28]

Although Isaac Canfield was a company director, and presumably dowsed all drilling locations, in 1883 he was working as a route agent and his father was a guard at the state prison. The United Oil Company absorbed Arkansas Valley Oil in 1887, and two years later, the Canfields went back into the oil business through the Triumph Oil Company, with Ira Canfield as president and Isaac Canfield general manager. They completed two successful oil wells in 1890, and another two in 1892.[29]

Ira Canfield became mayor of Florence, Colorado, in 1891. He died in 1895.[30]

The Spindletop gusher at Beaumont, Texas, in January 1901 inspired fortune-seekers to drill boldly in new places. Investors from Colorado Springs were ready to back Canfield in wildcatting, but where could he find the next Spindletop? Isaac Canfield decided to find another oil field—and he did.

Canfield may have based his new search on Ferdinand Hayden's 1877 geologic map of Colorado. The map showed that northeast of Boulder, Colorado, was a three- to five-mile-wide outcrop of the Colorado Group, a packet of geologic formations that includes the Pierre Shale, which is the reservoir rock in the Florence and Cañon City fields. It was a bold project, but starting in an area with geology similar to the Florence and Cañon City fields showed good sense. From there on, Canfield relied on his bobber.

Canfield formed various companies to drill his dowsed locations northeast of Boulder. The first well, the McKenzie, began drilling in July 1901. By early August, gas was coming out of the borehole, and the borehole started to fill with oil. Isaac Canfield's bobber had found a new oil field. His investors prepared to celebrate by shooting fireworks from the top of the derrick. The fireworks didn't go off, not because shooting fireworks around a well flowing gas was amazingly foolish, but because on August 22, the drillers lost the bailer down the hole. After three weeks failing to fish out the bailer, the hole caved in, and Canfield abandoned the well. They dismantled the derrick and reassembled it twenty-five feet away. Bad luck continued when the crew lost the drilling tool down hole, and the second borehole was abandoned. Canfield had to start drilling for the third time.

In the meantime, Canfield started drilling another well, the Arnold, half a mile north of the McKenzie. The Arnold hit oil, another victory for the Canfield bobber. An oil boom followed, with great hoopla and irrational optimism. More than a hundred oil companies were formed, and an oil exchange opened in Denver to trade oil company shares.

Promoters made up a fable that government geologist Ferdinand Hayden had predicted and mapped a great oil belt. In fact, Hayden had not even mentioned the possibility of oil. Yet the myth grew that Hayden had even driven stakes in the ground to mark the best places to drill. The Boulder Mining, Oil and Gas Company found an old post carved with the message "Center Oil Belt," which they assumed was placed there by Hayden.[31]

Canfield's success with his bobber inspired other companies to hire dowsers, and most of the wells drilled in the field were on the basis of dowsing. Other dowsers were less skillful. About a hundred wells were drilled in and around the Boulder field during the next four years, but only twenty-eight were productive.[32]

Drillers at Boulder were still careful not to tread on the toes of old oil field superstitions. Everything was ready to start drilling the Alamo well, but the driller shared the belief that it was bad luck to start drilling a well on a Friday, and so waited until the stroke of midnight turned the clock into Saturday before the crew began drilling.[33]

A. M. Hunter, the assistant Boulder County Recorder, invented an electrical apparatus he said was an infallible oil finder. One operator tore down his drilling derrick and reassembled it two hundred feet away, based on Hunter's machine. Hunter started using his machine to consult on oil well locations in Kansas and southeast Colorado.

Canfield cashed in his interest in the Boulder Oil Company and moved on to his next project, farther north around Fort Collins, Colorado. Promoters followed Canfield north and sold shares in competing companies. This time, however, his bobber bobbed wrong. He drilled three dry holes and gave up. The Fort Collins *Express* credited Canfield with a "game fight," but declared: "This ends the oil excitement in Fort Collins."[34]

Isaac Canfield moved to Western Colorado in 1904. Western Colorado newspapers called him "Captain" Canfield, although there was no previous indication of military experience.

W. C. Harding was a Detroit businessman who had invested in one of Canfield's dry holes near Fort Collins. He and Canfield leased 3,200 acres and drilled a wildcat well at Marysville, Ohio. Years previous, Canfield had passed through on a train and thought that the area looked like oil country. He dowsed a location with his bobber and started drilling, but the well was abandoned in April 1903 at a depth of 2,160 feet.[35]

Isaac Canfield later said that in 1903 he drilled forty-two oil wells in Ontario, Canada, with only one dry hole.[36]

First Oil Discovery in New Mexico—Almost

Prompted by crude oil floating in a water well, businessmen in Roswell, New Mexico, sent for Isaac Canfield, who dowsed a drilling location twelve miles east of Roswell and started drilling in March 1906. They found some oil in the San Andres Formation but discovered a lot more water than oil. The company shot the well with nitroglycerin and put it on pump in May 1907, hoping that the oil percentage in the water would increase. Instead, the oil percentage became even smaller, and after a week on pump, they abandoned the well, after producing about a hundred barrels of oil.[37]

Canfield's Roswell oil well was several years before the Brown well, drilled in 1909 at Dayton, New Mexico, about fifty miles south of Roswell, which is traditionally credited as New Mexico's first oil discovery. At Dayton, as with Canfield's well at Roswell, the oil was in the San Andres Formation, and mixed with large volumes of water. For years, the rancher at Dayton ran the production through large tanks to separate out the small amount of oil. Had Canfield dealt with the water, and produced his well for longer than a week, his well would likely have been regarded as the first oil discovery in New Mexico.[38]

Canfield had found enough oil to convince local people to put their money into the Southwestern Oil Company to drill for oil west of Roswell. It was a dry hole.[39]

Canfield then organized the Canfield Syndicate to drill for oil at De Beque, Colorado. The company drilled one hole there in 1904, but it was a failure.

After Isaac Canfield's string of failures in Colorado, Ohio, and New Mexico, he dropped from headlines for several years, but reentered the business with a high-profile series of successful wells that established his reputation as a legendary oil finder.

Oil operators were planning a test hole near the North Texas hamlet of Electra in 1909. After an initial dry hole, their driller talked them into sending for Isaac Canfield. He dowsed a location where the crew drilled and completed a well producing thirty barrels per day. It wasn't much, but it was the field discovery for the Electra field, and enough to attract other drillers. In July 1911, a well a few miles northeast started extraction at 1,500 barrels per day, which set off the Electra boom.[40]

In 1914, Canfield and a Texas cattleman named Tippit leased 420 acres in Okmulgee County, Oklahoma. Although their leasehold was surrounded by four dry holes, Canfield dowsed the tract, and they drilled eight producers into the Booch Sand, with no dry holes. They then sold the property for $300,000.[41]

Fig. 10. The surface equipment for Canfield's discovery well for the Boulder oil field was preserved as a historic monument as the land developed around it. (Photo by the author.)

In 1919, Isaac Canfield lent his name to oil promotions, and in turn his clients puffed him up to legendary proportions in their stock-selling advertisements. Newspapers referred to him as "Colonel" Canfield, the famous geologist.[42] He was also falsely described as the discoverer of the Petrolia and Ranger fields in Texas and the Muskogee field in Oklahoma.

Isaac Canfield died in 1924, aged eighty-five, as one of the most successful oil dowsers in history.[43] He worked in the oil business for more than sixty years, from Pennsylvania before the Civil War to Texas during the Jazz Age. He drilled the first deep oil well in Colorado and discovered the Florence and Boulder oil fields, the second and third oil fields in the state. He then found the first—although noncommercial—oil in New Mexico and dowsed the Electra field discovery well in Texas.

His bobber was not infallible. Contrary to his bragging, he drilled too many dry holes to prove the validity of oil dowsing. But by the end of his oil-dowsing career, he had a record of discoveries that would be the envy of nearly any geologist.

7

OIL DOWSING IN THE AGE OF GEOLOGY

THE RISE OF THE GEOLOGISTS

For the first forty years of the oil business, well locations were based on nearly everything except geology. Oil historians Harold F. Williamson and Arnold R. Daum noted of the early oil drilling in Pennsylvania: "The services of professional geologists were at once almost wholly unavailable and useless."[1]

The first oil operators to hire a permanent geologist were the Nobel brothers, who hired Swedish geologist Hjalmar Sjögren in 1885 to increase their Russian oil production. In the late 1800s, geologists were employed in the oil fields of Alsace, France, and of Galicia (now in southern Poland).[2]

European oil companies spread geologists around the world. Shell Oil hired two geologists—an Italian and a Swiss—in 1898 to reverse declining production in the Dutch East Indies (now Indonesia). European companies even sent geologists to the United States. The Italian geologist Cesare Porro, working for the Dutch company *Petroleum Maatschappij Salt Creek*, chose the location of the discovery well for the giant Salt Creek field in Wyoming. One of the first oil geologists working full-time in Oklahoma was Hans Hirschi, a Swiss who arrived in 1911 for the *Union des Pétroles d'Oklahoma*.[3] Edward Doheny drilled the first commercial oil well in Mexico in 1904 at a location chosen by maverick Mexican geologist Ezequiel Ordóñez—other operators in Mexico followed Doheny's example and hired geologists.

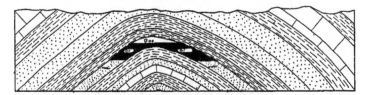

Fig. 11. An anticline is a ridge-shaped fold of stratified rock in which the strata slope downward from the crest. In this image, the oil and gas are found underneath a layer of shale (the layers represented by horizontal lines) and sandstone (the layers represented by dots).

California led the rest of the US in employing oil geologists. In 1898 the Union Oil Company, then the largest oil producer in California, became one of the first American oil companies with a geology department. Early California oil wells were drilled at seeps, places where oil could be seen oozing out of the ground. As drillers ran out of seeps, they hired geologists to choose well locations. From 1875 to 1902, twenty-eight California oil fields were discovered, all based on seeps. In the decade 1903–1912, twelve fields in California, six of them giant fields, were found by geology, versus two found by seeps. By 1908, all the major oil companies in California employed geologists. Surface geology was also proving itself in Wyoming, but not further east.[4]

Despite success overseas and in California, oil operators east of the Rockies doubted that geologists could contribute anything of value. This was partly due to the jealously guarded autonomy of oil operators. From the start of the American oil industry, its heroes were the wildcatters: fiercely independent, unconventional, ready to gamble it all on the next well. Ida Tarbell celebrated early small operators as "a triumph of individualism."[5] The Standard Oil refinery monopoly had squeezed the independent operators in the eastern fields, but in the Gulf Coast and midcontinent regions, due partly to populist politics hostile to Standard Oil—and partly to the oil business expanding too rapidly for Standard Oil to control—the iconic wildcatter emerged stronger than ever in the early twentieth century. Wildcatters were jealous of their independence and resisted letting anyone, dowser or geologist, tell them where to drill.

When an executive of Standard Oil of Ohio went west to take charge of Standard Oil of California in 1911, he brought the disdain for geology, and immediately fired all but one of the geologists. After a string of expensive dry holes, however, he hired another geological staff.[6]

Entering the second decade of the twentieth century, geologists were widely employed in many of the world's major oil-producing regions, but not in the United States east of the Rockies, which produced a third of the world's oil.

The Cushing, Oklahoma, gusher in March 1913 was a gift to geologists because drillers at Cushing learned that if they weren't on an anticline, they drilled nothing but dry holes. The wells at Cushing produced more than a quarter of all the oil extracted in America in 1915.[7]

Geologists had been arguing over the anticlinal theory since 1885. The trouble was that there were anticlines barren of oil, and there were oil fields unrelated to anticlines.[8] But after Cushing, oil operators realized that although not every anticline held oil, drilling anticlines was much better than drilling at random.

The people who could find anticlines were geologists. With a plane table and alidade survey instrument, a Brunton compass, and a few rock outcrops, a pair of geologists could make a reasonable map of geologic structure before a lease was bought or a well was drilled.

Cushing was not found through geology, but geologists for the Gypsy, Hill, and McMan companies showed that field development at Cushing was more efficient using geology.[9]

Empire Gas and Fuel hired young geologist Everett Carpenter, but the company would not spend $100 for an alidade so Carpenter bought one anyway, for which Empire reprimanded him. He and his alidade mapped a structure in Kansas. After great effort, Carpenter convinced the company to lease and drill the structure, and in 1914 they discovered the giant Augusta oil field. The following year Carpenter discovered the nearby giant Eldorado oil field. The discoveries leapfrogged the small company Empire Gas and its parent, Cities Service, into a major corporation in the Midcontinent. Oil operators had to admit that geologists were on to something, and companies began hiring geologists in earnest. Texaco hired its first geologist in 1913, as did Gypsy Oil, the exploration arm of Gulf.[10]

Geologists found other anticlinal oil fields, prompting more oil companies to hire geologists. In the period 1906 to 1910 only one giant field in the US (out of five) had been found using geology, but between 1916 and 1920 geologists discovered thirteen giant US fields (out of twenty-three).

Lurking behind the question of finding oil was the mostly unspoken clash of social class posed by college-educated geologists invading a proudly working-class industry. Many operators had worked their way up from the drill floor and had little good to say about geologists who considered it undignified to get dirty.

Wallace Pratt recalled that many early geologists had a condescending attitude. When he started at Humble in 1916, the chief geologist came to work in striped pants and claw-hammer tailed coat. A geologist in 1914 contrasted the older, academic geologist as one "who wasted little time among the briars and rush," compared to younger geologists who were not above field work. Geologists who entered the industry after 1913 were willing to get dirty, sweaty, and bug-bit, fighting through brambles to the next outcrop—or if they weren't, they didn't stay long in the industry. Pratt thought that the hiring of geologists was due more to the changed attitudes of geologists than the changed attitude of the industry.[11]

By 1915, the *Oil & Gas Journal* noted that all but one of the big oil companies in Oklahoma had geologists on staff. Prejudice against geology died hard. The first geologist hired by Carter Oil was told by his boss: "I don't think you will do us any good and I will damn well see that you don't do us any harm." In 1918, the head of an oil company in Oklahoma and Kansas hired his first geologist but kept him on the records under another job classification, to avoid ridicule. The head of another Oklahoma company hired a geologist in 1920, but paid for him out of his own pocket, to avoid criticism for squandering company money.[12]

By the early 1920s, science had won the battle. It was acknowledged that crude oil could be found based on scientific principles. But what was science? For many people, science was incomprehensible mumbo-jumbo, and anyone who could spout scientific-sounding patter was a scientist. Enter the doodlebug.

Geologists and Dowsers

The rise of geologists did not coincide with a decline of oil dowsers. On the contrary, the decades of the 1910s and 1920s, which saw the establishment of oil geologists, also saw a large increase in oil dowsing, based upon mentions in the news media. Geologist Wallace Pratt observed that many oil drillers lumped geology and dowsing together.[13] Oil men faced what seemed to them to be two bad choices: geologists or dowsers/doodlebuggers. After some dry holes in Montana, oil promoter E. G. Lewis fired his geologists and hired a doodlebugger. After more dry holes, Lewis fired his doodlebugger.[14]

As geologists moved up into management, they used their influence to keep dowsers out. Oil dowsers have clung to a small share of drilled wells by great persistence and cultivating clients among small independent operators. Oil dowsers are still active in the United States, but their number is dwindling.

It is often found in dowsing literature that the practice could almost eliminate dry holes. So many dowsers boast 90 percent success that the statistic is almost a cliché. It is the rare oil dowser who admits to a success rate of less than 75 percent.

Geologists aren't perfect, and oil dowsers don't have to be perfect, either: they just have to be better than geologists. Dowsers often claim that 90 percent of well locations chosen by geologists are dry holes. But this is false for wells in general, as the proportion of wells drilled in the United States that produce oil and gas has never dipped below 57 percent in any year since the data were first reported in 1908.[15]

Oil and gas wells can be categorized as exploratory (wildcat) wells or development wells. The success rate of wildcat wells often hovered around 10 percent, though the rate has changed as years passed and is different for different areas. For the period 1935–1989, one in eight wildcats was successful. Development wells, those within already-discovered fields, have much higher success rates, which may exceed 90 percent. In evaluating dowsing, the type of well is crucial.

Dowsers promise a very high rate of success in oil drilling. But can the dowsers deliver?

A BOTTLE OF CRUDE OIL HANGING BY A STRING

Fourteen-year-old William Henry Zachary was plowing the family farm near Lockhart, Texas, when he stopped because of a strange tingling in his hands. His mother remembered that her father had used tingling in his hands to find underground water. The Zachary family dug a well at the spot and found water. Soon Henry was locating water wells all over Caldwell County.[16]

For $100 and a small share of the oil revenue, Henry's parents let Lockhart businessman Elliott take their son to Beaumont, Texas, to test his tingling hands over the new Spindletop oil field. Henry Zachary felt the same sort of sensation at the oil field, and based on Zachary's tinglings, Elliott leased land and drilled some successful oil wells.

Henry spent the summer of 1902 with family friends in Pueblo, Colorado. Prospectors took him to Cripple Creek and the San Juan Mountains, where the tinglings convinced him that he was also sensitive to ore deposits. The

Fig. 12. Newspaper cartoon of Henry Zachary, with the caption: "Wichita Falls – The Luke Wilson Oil Company has hired Henry Zachary, an official 'oil smeller,' to smell for oil on their land." (*The Evening News* (Harrisburg, PA), April 19, 1917.)

prospectors staked claims, but the results are unknown. The following year, Lockhart businessman John Shoaf took him to find gold at Johannesburg, California. To augment his tingling hand, Zachary used a piece of whalebone with a gold coin attached to the tip.[17] In 1904, mine owners in Alaska contracted with him to search for gold. They agreed to pay Zachary $100,000 spanning ten years, but the effort seems to have ended after the first summer, suggesting that he had little success.

In 1905, Lockhart businessmen formed the Magnetic Oil Company to explore for oil with Zachary's hands, which tingled over a spot north of Lockhart. No results were reported.[18]

In 1906, nineteen-year-old Henry Zachary left for Nome, Alaska, again in the company of John Shoaf. They said they would not return until they had "struck it rich," but they returned at the end of the summer, with no sign of wealth.[19]

Zachary moved to San Antonio about 1909. He dowsed oil locations east of the Humble field, and his client drilled one of the locations to 2,200 feet before they abandoned both the well and their faith in Henry Zachary. A few years later, however, other drillers reportedly drilled prolific oil wells 1,000 feet deeper. In 1914, Zachary dowsed ten oil prospects for oil man E. O. Burton. Twenty-two years later, Burton told a reporter that seven of the ten prospects had become oil fields, the eighth had just hit oil, and the remaining two were undrilled.[20]

By 1915, Zachary was using a bottle of crude oil hanging by a silk cord. The bottle was closed with a cork stopper, with one end of the silk cord immersed in the crude oil. He said that he could feel a peculiar vibration transmitted up the silk string when he was over oil.[21] Zachary dowsed oil near Pilot Knob south of Austin, Texas, in 1915, and predicted a 7,000-barrel per day gusher. His Pilot Knob location was put to the drill bit but was a dry hole. Another high-profile dry hole was a wildcat test dowsed by Zachary in Russell County in 1924.

In 1917, Zachary advertised as an "oil locator" in classified ads of the *Oil & Gas Journal*. He dowsed wells for the Uncle Luke Wilson Oil Company and left that client highly impressed. That same year, he went to Wichita, Kansas, and won a loyal clientele. He established a second home in Wichita, and for the next decade split his time between Wichita and San Antonio. The Wichita *Daily Eagle* noted that oil driller J. E. Atkins would not drill a well unless Zachary had approved the location. In 1924, the Hutchinson *News* wrote of Wichita, "There are oil men in that section who wouldn't 'shoot' a well without Zachary's advi[c]e."[22]

Henry Zachary organized the Globe Petroleum Trust in 1920 to try again to find oil in Caldwell County. The trust drilled a location east of Lockhart, where Zachary had dowsed an oil reservoir at 900 feet. He claimed to be able to hit oil in eight out of ten wildcat wells, and projected returns of fifty dollars or more for every dollar invested. Unfortunately, it proved another dry hole.[23]

Henry Zachary never again achieved the acclaim he enjoyed in Wichita in the 1920s, but his classified ads in the San Antonio newspapers seemed to bring him steady work. While keeping his primary residence in San Antonio, he followed the oil rigs. He opened an office in Laredo in the 1930s, relocated for a few years to Colorado City in West Texas, then in 1942 moved his remote office to Fort Worth. In 1950, Henry Zachary moved to San Angelo, Texas. William Henry Zachary retired in 1952 and died in 1961.

RAISING HELL AND DOWSING FOR OIL

A. C. Townley was a politician, not a dowser. But he used his popularity to promote oil drilling guided by dowsers. Like other celebrity oil promoters, Townley never found oil, and wasted the money entrusted to him.

Arthur C. Townley grew up in Minnesota and taught school for a couple of years, then farmed in North Dakota before moving to the eastern plains of Colorado to raise wheat. He failed at wheat and returned to North Dakota in 1907 to raise flax. After five years he was "the flax king of the Northwest." But an early frost in 1912, coupled with a fall in the flax price, bankrupted him. An embittered A. C. Townley switched from budding capitalist to socialist organizer and ran unsuccessfully as a Socialist Party candidate for the state legislature in 1914.[24]

Townley saw that the Socialist Party was going nowhere, so in 1915 he organized the Nonpartisan League to press for socialist ideas through the existing Democratic and Republican parties. He advocated state-owned grain elevators, flour mills, meat-packing plants, banks, and crop insurance. The Nonpartisan League was only a year old in 1916 when league-endorsed candidates won the governorship of North Dakota and a majority of the lower house of the legislature. In 1918, its candidates won all statewide offices and more than two-thirds of both houses of the legislature.[25]

Townley was a compelling orator and an organizing genius, but he had powerful opposition. He was accused of being pro-German during World War I, was convicted in 1921 of conspiring to interfere with the military draft and the sale of Liberty Bonds, and served ninety days in jail. When the Nonpartisan League lost power in 1922, Townley left the state. In 1924, A. C. Townley told a newspaperman, "Politics is bunk."[26] By March 1925, he was in Texas and Kansas, selling fractional interests in oil wells.

The people of Robinson, North Dakota, became excited when a water well in town was discovered to have a layer of light oil floating on the water—oil so light that it resembled gasoline. In fact, it was gasoline. State geologist A. G. Leonard wrote that chemists had even identified the brand as Red Crown. He wrote: "The oil found in the water wells has evidently been poured in from above."[27] But he was widely ignored. Oil companies crowded into Robinson and leased land all around. Among the out-of-state oil men was the familiar figure of A. C. Townley, who returned to North Dakota in August 1925 to drill for oil.

Rumors reached the North Dakota Industrial Commission that Townley was selling oil stock. Townley explained to reporters that he wasn't selling shares. Instead, he was raising money for oil drilling by borrowing the money from small investors, for which the investors received unsecured IOUs bearing

9 percent interest. The North Dakota Industrial Commission had no jurisdiction concerning an individual borrowing money.[28]

In October 1925, Townley announced that he would drill for oil a mile from the Robinson gasoline well. He was sure that his location was atop an oil field because his oil dowser, Henry from Texas, had detected oil there using a bottle of oil hanging from a silken cord. The dowser's last name is not mentioned in news reports, but it was likely the well-known oil dowser Henry Zachary.

Townley was a compelling salesman with a large reservoir of goodwill in North Dakota, and he knew very little about drilling for oil. Such ignorance was a drawback for finding oil, but it freed the imagination wonderfully for raising money.

Townley started drilling in December 1925 and halted in the spring of 1926. In May 1927, he announced that his well had struck oil, and exhibited a bottle of crude oil dipped from his well. He said that he had withheld the announcement for more than a year to pick up additional leases. But Townley made no effort to produce oil from his well. Instead, he used the publicity to raise more money, and drill more wells.[29]

Townley's second well started drilling near Nesson, North Dakota, in early 1928 while he was building three more drilling rigs scattered around the state at Ray, New England, and Valley City. Townley ran out of money before he could finish any more wells. Creditors foreclosed on his drilling equipment in December 1929. The *Bismarck Tribune* estimated that Townley had raised— and lost—between half a million and a million dollars from his followers.[30]

In April 1951, Amerada Petroleum made the first oil discovery in North Dakota, on the Clarence Iverson farm near Tioga. Other companies rushed in. A year after the oil discovery, with North Dakota in the middle of an oil boom, A. C. Townley returned to North Dakota and announced, "I am calling the old crowd together." He explained, "I think we ought to finish the job."[31]

Townley's meeting on June 3, 1952, was at the Iverson farm, the site of Amerada's oil discovery. About two hundred attended, mostly farmers from the area. Townley brought back his dowser: Henry from Texas, who still used his familiar small jar of petroleum hanging by a silk cord. Townley was sure that this time the doodlebug would find oil.[32]

Townley told the landowners that he wanted them to sign over half their mineral rights to him. In return, Townley would pay all the expenses, and pay the mineral owners half the profits. At the mention of signing over mineral rights to Townley, his audience started to walk out. By the time Townley finished, less than two dozen remained. Townley's plan found no takers. His host remembered the meeting Townley had held on the same farm twenty-seven years before. Clarence Iverson had been fourteen years old when his father, a Norwegian immigrant, invested $500 in Townley's oil dowsing and drilling, and lost it all.[33] Townley, with his oil dowser and his trust-me-with-your-mineral-rights scheme, seemed as outdated as his old prairie socialism.

Arthur Townley took a job selling insurance. He ran for the US Senate in 1956 as an independent, endorsing Republican President Dwight Eisenhower, and decried what he said was communist influence in Washington. He now said, "Socialism doesn't work." Townley drew 937 votes out of 244,161 cast. He ran for North Dakota's other Senate seat in 1958 and got 1,700 votes. By the time he died in a car crash in 1959 at age seventy-nine, A. C. Townley, his flaming oratory, and his attempts to find oil, were all nearly forgotten.[34]

ALOIS SWOBODA AND THE SUPERIOR LIFE

The Austrian immigrant Alois Swoboda was America's original muscle-building guru. His advertisements promised not only a muscular physique but total health inside and out, perpetual youth, and cures for everything from malaria to bad breath. He taught a philosophy of positive thinking, immodestly named "Swoboda-ism." His large advertisements challenged: "Why be satisfied with only a half?" and "Why live an inferior life?"[35]

By the early 1920s, Swoboda had a large following, and he expanded his teachings into finance. In 1923, he branched out into mining and oil, bragging that he had a "money sense" that enabled him to find ways to make his Swobodians rich. If they would only invest in Swoboda's oil wells, their money would multiply unbelievably.[36] As usual, the *Engineering and Mining Journal* could be counted on for a blunt assessment: "We know of no more painful and nauseating reading than the come-on letters of Alois P. Swoboda."[37]

On a train from Montana to Salt Lake City in 1927, Swoboda met John Dahlgren, a farm hand from North Dakota. Just south of Pocatello, Idaho, Dahlgren remarked that he could feel the pull of a huge oil field to the south. When the train reached Ogden, Utah, Dahlgren said that the pull was from the west, placing the oil northeast of the Great Salt Lake.

Swoboda had considered investing in an oil company preparing to drill northeast of the lake. The following day, Swoboda brought Dahlgren along to inspect the oil leases of the Lenora Mining and Milling Company. They walked the ground until Dahlgren stopped and announced that he was standing on the place to drill. It was the same spot chosen by the Lenora geologists, which impressed Swoboda enough to invest in the Lenora company, and to buy one of their leases and drill it himself. Swoboda and the Lenora company both began drilling in November 1928. Both wells were abandoned as dry holes by the end of 1930.[38]

John Dahlgren said that he could also sense gold and silver. Dahlgren could go into a trance and converse with the dead. According to Swoboda, one evening Dahlgren had a long conversation with Francis Bacon. But it was John Dahlgren's oil-finding ability that Alois Swoboda introduced to the world in 1928.

Swoboda and Dahlgren went to Townsend, Montana, where geologists had chosen three drilling locations. Dahlgren walked around with a bottle of crude oil in his hand until he found a location ten feet from one of the spots chosen by geologists. Dahlgren continued until he found two more spots, each within sixty feet of a geological location, but all three locations were dry holes. Swoboda leased 70,000 acres and drilled a dry hole near Edgeley, North Dakota.[39]

In June 1930, after eighteen months of selling stock in John Dahlgren's oil-sensing, a New York court ordered Swoboda to stop selling shares in that state.[40] Blocked from selling shares in mines and oil wells, Swoboda shifted the hunt. In December 1930, he announced that he and Dahlgren would search Death Valley for the Lost Horse Mine. Nothing more was heard of the search, and nothing was heard of Dahlgren after 1930. Alois Swoboda died in 1938 at age sixty-five.

HENRY GROSS: AMERICA'S MOST FAMOUS DOWSER

According to the dowsing literature, Henry Gross's most notable achievements were: first, that he dowsed the first freshwater wells on the island of Bermuda, where scientists said there was no fresh groundwater; and second, that he correctly dowsed fifteen out of seventeen undrilled wildcat oil-well locations in Kansas. Sadly, Henry Gross did neither.

Gross was a game warden who dowsed to help his Biddleford, Maine, neighbors choose sites for water wells. He was like many small-town dowsers until at age fifty-six, he starred in the book *Henry Gross and His Dowsing Rod* by

Kenneth Roberts. Roberts was a Pulitzer Prize–winning writer of best-selling historical novels when Henry Gross located a water well on Roberts's estate at Kennebunkport, Maine, in 1947, and impressed Roberts with his confident ease in handling a forked twig. With a world-class dowser such as Henry Gross, Kenneth Roberts thought that he could irrefutably prove the case for dowsing. Roberts became Gross's business manager and recounted their experiences in three books: *Henry Gross and His Dowsing Rod* (1951), *The Seventh Sense* (1953), and *Water Unlimited* (1957).

Henry Gross is commonly celebrated for map-dowsing from Maine the location of the first freshwater wells on the island of Bermuda—except he didn't.

In Maine, Henry Gross held his dowsing rod over a map of Bermuda while Kenneth Roberts slid a ruler back and forth over the map, marking the locations where the rod dipped. Roberts and Gross flew to Bermuda in late 1949. Roberts chose Bermuda because he had the mistaken idea that fresh groundwater had never been found on the island. A 1913 geological report noted that there was fresh groundwater on Bermuda, but the wells were "shunned for sanitary reason" as Bermuda groundwater is commonly contaminated with bacteria and nitrate, and freshwater wells tend to become brackish or salty after being pumped.[41]

Two of Gross's dowsed wells found fresh water; the third found brackish water. Later investigations revealed that Bermuda was underlain by five bodies of fresh water, the largest of which covers more than two square miles.[42] When Henry Gross chose locations within freshwater bodies (twice), he found fresh water. When he chose a location outside the freshwater bodies, he got salty water. The freshwater areas are now well delineated by hundreds of wells, and there is no doubt about this. Henry Gross made no out-of-ordinary water discoveries on Bermuda, and made no difference to the water supply of the island.

Gross was far from an infallible water dowser. Some of his more notable failures are: Laurel, Maryland (Johns Hopkins Applied Physics Laboratory); Eleuthera Island, Bahamas; Albany, New York; Salem, Massachusetts (Sylvania Electric Company); Tikal, Guatemala (Philadelphia Museum); and Libya (US foreign aid program).[43]

Oil Dowsing

Kenneth Roberts was hesitant to have Gross dowse oil, but it offered greater visibility and higher income. Henry Gross started oil dowsing in 1952 when some Canadians compared Henry Gross's oil dowsing with the findings of four other dowsers. One of the other dowsers agreed with Gross. The other three disagreed, so Kenneth Roberts dismissed them as "obviously inexperienced," and complained that the study was "supervised by an incompetent observer," presumably meaning that Roberts didn't like the results.

Gross dowsed an area in Alaska where he said there was a large undiscovered oil field. He received a block of shares in the oil company, but apparently nothing of value was found.

Rudolph Johnson drilled for psychic Edgar Cayce's "Mother Pool" of oil in San Saba County, Texas, but the drill went down to 3,500 feet without finding oil. Gross map-dowsed the drill site from Maine and said that the oil had once been at 2,600 feet, as Cayce prophesized, but had moved downward about 1,800 years ago and was now at 5,200 feet. Johnson could not afford to drill that deep, so he asked Henry Gross to find bypassed oil in existing wells.

Gross told Johnson that two old oil wells in Coleman County, Texas, had unproduced oil at 1,130 feet. The well logs confirmed a sand layer at 1,137 feet. Johnson bought the old wells and produced oil from the sand. The results were mediocre, and Gross's share of the oil was less than nine dollars per month. Roberts complained: "Thus our total return for two years of heavy letter-writing and exhausting map work returned us a total of $368.79." Welcome to the oil business, Mr. Roberts, where you are not rewarded for "heavy letter-writing."

Some Hollywood investors flew Henry Gross to California, where they ignored the known oil areas of California and drove Gross around the southern California deserts, where no oil has been found. Gross dowsed fifteen oil fields: three near Lancaster, six around Owens Lake, one near Darwin, one near Randsburg, and four in the Panamint Valley. Someone riding with Gross marked the locations on a pair of road maps. None of Gross's locations appears to have been drilled. The depths given Gross by his dowsing rod were sometimes greater than the thickness of sediments, putting oil reservoirs up to several thousand feet into igneous or metamorphic basement—a highly unlikely place to finding oil.

The investors still wanted to know if Gross could locate oil, so they sent Henry Gross to Texas to dowse well locations not yet drilled; Roberts called the locations "wildcats," meaning that they were away from known oil fields. Of the Texas locations dowsed by Gross as having oil, fifteen were later drilled, of which seven became oil wells. Roberts noted that seven out of fifteen was much better than the expected success rate for wildcat wells. He did not identify the wells, so the results cannot be confirmed.

Superficially, Henry Gross's most impressive oil-dowsing was correctly dowsing fifteen out of seventeen undrilled wildcat oil well locations in Kansas, which had only a 2 percent probability of happening by chance—or so wrote Kenneth Roberts. The truth is less impressive.

Two Hollywood investors had backed a drilling program by the Imperial Oil Company of Kansas, and the company arranged for Gross to dowse locations for thirty-six future oil tests. Kenneth Roberts emphasized in *Water Unlimited* that all thirty-six drill sites were wildcat locations. But just five pages further in the book a letter from the president of the Imperial Oil Company of Kansas is reproduced listing eight of the thirty-six locations as not wildcats. In addition, seven of the thirty-six locations had already been drilled before Henry Gross dowsed the spots. This should have been known by Roberts.

Of the thirty-six prospects, seventeen were fully drilled by the end of the summer. According to Roberts, Gross was correct in thirteen of the results, and incorrect in two; the other two results were judged (by Roberts) to be "probably" correct, and "may be" correct. Roberts crowed: "[N]o oil geologist had ever come within a million miles of his accuracy."[44]

Robert Williams, the president of the Imperial Oil Company of Kansas, was unimpressed, and wrote to Roberts that they should consider only wildcat locations drilled after the dowsing, and more tests were needed. He noted that as long as you predict that wildcat wells will be dry holes, you will be correct most of the time. Roberts's reaction was predictable: he wrote on the letter in his characteristic red pencil: "This is plain bullshit—The old run-around = double talk!"[45]

Removing the wells drilled before Gross arrived leaves nine tests: six wildcat and three development wells. Of the wildcats, one discovered oil, and five were dry holes; the three development wells resulted in two oil wells and one dry hole. If the driller had followed Gross's advice, he would have drilled two of the wildcats, making one discovery and one dry hole. Gross would have drilled two development wells, both of which would have been oil wells. Overall, rather than drill nine holes to make three oil wells, following Gross's dowsing would have drilled only four wells to discover the same three oil wells. It was a very good result for Gross, but not nearly as good as represented by Roberts.

Gross's oil-well dowsing in Kansas convinced the investors, who hired Gross to evaluate six wildcat oil well locations in Colorado. The locations were all high-risk wildcat wells, scattered across the eastern plains of Colorado where very few wells had been drilled, and all dry holes. Gross predicted that four of the wells would strike oil at depths of between 3,600 and 4,200 feet. To hit four out of six wildcat wells of this nature would have been a stunningly good result. The investors did not depend on Gross's depth estimates and drilled each well deeper than the oil depths predicted by Gross, down to test the D Sandstone and the J Sandstone, both prolific oil producers in other parts of the Denver Basin.

The wells were drilled in the summer of 1956, and all six were dry holes. Roberts insisted that Gross had correctly detected oil but erred as to the depth. The four wells where Gross had dowsed oil had drilled between 410 to 740 feet past the depths where Gross had predicted oil. The investors lost interest in Henry Gross.

Les Egbert, who owned land in Breathitt County, Kentucky, had Henry Gross map-dowse the properties for oil in September 1958. Gross reported that two of the Egbert tracts contained oil. He followed up with field dowsing and returned for more oil dowsing in 1959 and 1962.

According to psychic researcher Stephan Schwartz, Gross predicted that six wells would be profitable (defined by Schwartz as producing at least one barrel per day per hundred feet of depth), and three would be dry holes. Schwartz wrote that Gross was correct on all three dry holes, and right on four out of the six locations where he predicted oil, for a total of seven correct out of nine in an area without any previous oil wells, and not regarded as prospective by geologists.

Gross predicted three oil sands, at 1,300, 1,600 and 3,200 feet, which would yield, respectively, seventy-six, 105, and 210 barrels of oil per day. Egbert found an oil sand at about 1,000, a bit shallower than Gross predicted, and yielding much less oil than the forecasted seventy-six barrels per day. Schwartz acknowledged that Gross had greatly overestimated production rates but excused him on the theory that although Gross predicted the oil pumping rate in barrels per day, he really gave the rates in gallons per day (at forty-two gallons per barrel). This jiggling of the numbers after the fact is a variation of the infamous Texas sharpshooter fallacy.[46]

Schwartz's definition of a successful well, one barrel per day per hundred feet of depth—a rule he presumably got from Les Egbert—is a very low threshold for success. No wonder geologists did not regard the area favorably. Even at that low definition of success, at least one of the wells that Schwartz wrote was a commercial oil producer, in reality only made six barrels per day from a 1,091-foot well—a commercial failure.[47] Results of some wells could not be confirmed, but Schwartz seemed overly eager to validate Gross's oil dowsing.

In 1963 and again in 1965, Egbert drilled holes within Gross's field outline well past the 1,600-foot depth of Gross's second oil sand but found nothing.[48] Gross correctly predicted that there would be oil at a shallow depth in an area not known to have oil, but he greatly over-predicted oil production rates, and predicted a second, deeper oil sand that was not there.

Good-natured, grandfatherly Henry Gross became the face of dowsing in America. When he died in 1979, the magazine *The American Dowser* hailed him as "the prince of dowsers."[49] The dowsing of Henry Gross is as well documented as that of any dowser, but his record is too mixed to conclude that Gross's dowsing rod reacted to more than his own subconscious.

DOWSING NUTT AND OIL-DOWSING'S LAST HURRAH

"I've been trying to tell people for 35 years that there is oil here."

—Roy Nutt[50]

Noel J. Monk promised his wife an oil well. He was a salesman, and they lived on a farm about thirty miles north of Columbus, Ohio. As a young man he had worked on oil rigs around Hobbs, New Mexico, and drilling for oil was no mystery to him. In 1957, he hired a driller who bored down to 3,800 feet without finding oil. The result was not surprising, because sixty-three wells had been drilled into Morrow County looking for oil, all of which were dry holes.

Monk abandoned his oil search until he met Roy Nutt, a farmer and well-known water dowser in Central Ohio. Nutt was convinced that there was lots of oil beneath Morrow County, and that his dowsing rod could tell him where. He loved to tell people about the oil, sometimes more than people wanted to listen. But Noel Monk listened when Nutt told him that there was oil beneath his farm and thought that Nutt's dowsing rod was worth a test, and besides, Monk had promised his wife that oil well. So he, Roy Nutt, and thirty-five of Monk's neighbors invested in the Monk Oil Company, and Roy Nutt dowsed a place to drill on Monk's farm, 2,000 feet southeast of the first hole. A week before Christmas 1959, the drill reached 3,731 feet, and the well started flowing 240 barrels of oil per day. Monk trusted Nutt's oil-dowsing so much that he had already installed oil flow lines and storage tanks while the well was drilling.[51]

Monk kept the discovery quiet, and over the holidays signed fifteen oil leases covering 1,096 acres—wherever Roy Nutt's dowsing rod told him to put a peg in the ground. Roy Nutt was seventy-four when he made his first and last oil discovery. Monk tried to give him all the credit, but when a newspaper cropped Nutt out of a photo of him and Monk, he blamed Monk, and the partners broke up.[52]

Nutt's discovery attracted other companies to drill nearby, but without much success. Monk moved the drill rig across the road to drill on the property of his neighbor, who was an investor in Monk Oil. It was a dry hole.

By mid-1961, oil companies were losing interest until the United Producing Company drilled the Orrie Myers well. It came in during June 1961, capable of flowing 2,000 barrels of oil per day, and the pre-Trenton play opened into a true oil boom. Thousands of wells drilled to the play horizon found dozens of small fields scattered beneath southwest Morrow County.

The oil editor of Ohio's Mansfield *News Journal* cited an unnamed source familiar with the drilling that a "high percentage" of the well locations— especially those by smaller oil companies—were placed by dowsers, although the companies kept it quiet. A geologist remembered that every day, three or four oil dowsers could be seen walking through the woods, and one or two panel trucks—presumably carrying elaborate doodlebugs—roamed the back roads.[53] A woman from Oklahoma said she could feel vibrations from underground oil, and drove back roads in a pickup truck, stopping to give drillers unasked-for advice about moving their rigs a few feet this way or that. Dowser "old man Mosher" traced a river of oil twenty feet wide across the countryside and told a driller that he needed to move his location twelve feet south if he wanted to hit the oil. If the observation was accurate that oil-dowsing was commonly used in the Central Ohio oil boom of the 1960s, it was probably the last such place in the United States. It was oil-dowsing's last great hurrah, though a silent one.

PAUL CLEMENT BROWN

"At every location I dowsed, commercially worthwhile quantities of oil or gas were found."

—Paul Clement Brown[54]

Paul Clement Brown is famous among dowsers for predicting the results of a series of thirty-five oil wells—and successfully predicting all thirty-five. He never sought fame and seems to have avoided media interviews and publicity.

Paul Clement Brown grew up in Oklahoma and graduated from the University of Texas and MIT. After some time in Boston, he moved to Southern California, worked as an electrical engineer, then opened a radio repair shop. Between replacing blown-out vacuum tubes, he experimented on ways to find oil. By the 1950s, he had perfected his technique.

Brown looked for petroleum using a small oil-filled vial on the end of a string, an empty gallon milk carton, and a stopwatch. He would hang the vial inside the milk carton to prevent breezes from swaying the vial, and then start the stopwatch. The stopwatch had a special face graduated in feet instead of seconds (ninety feet per minute). When the stopwatch dial reached the depth of the top of the oil horizon, the vial would begin to swing in a counterclockwise circle; then when the stopwatch reached the base of the oil-bearing zone, the vial would stop swinging. Brown gauged the productivity of the oil zone by the intensity of the swings.[55]

According to Brown, wildcatter Lee Davis recommended him to J. K. Wadley. Wadley had Brown dowse a wildcat location he was about to drill in the Cascade-Aliso area of Los Angeles County, California. Brown dowsed the well, and the Teater-Wadley #1 found oil as predicted. It was completed in October 1954. Wadley sought Brown's advice again, but this time Brown disagreed with Wadley's geologists, and concluded that the well in White County, Arkansas, would be a dry hole. Again, Brown was right.

Brown told Christopher Bird that he dowsed a large uranium deposit near Grants, New Mexico, after J. K. Wadley drilled a dry hole for oil at the location. Brown related that Wadley had sent him to the old dry hole to see if there was a good oil location elsewhere on Wadley's lease. Brown told him that there was no oil on the lease, but that it was rich in uranium. But Wadley's authorized biography does not mention Brown, and credits the uranium discovery not to dowsing, but to government scientists who tested samples of a uranium-bearing sand that Wadley had drilled through in 1952 while looking for oil.[56]

Brown said that he convinced J. K. Wadley to drill the well that found the northwest extension of the Dominguez oil field in Los Angeles. However, Wadley's authorized biography does not mention dowsing or Brown, and credits the discovery well to an unnamed engineer.

A number of dowsing writers have emphasized that it was the oil wells that Brown dowsed at the Dominguez field that started Occidental Oil Company.[57] Occidental Oil Company had been around since 1920 and wanted to expand, so the company bought Wadley's Dominguez wells in 1956 for $1,750,000. This was after Brown's locations had already been drilled. If Occidental had not bought Wadley's oil wells, the company no doubt would have bought some other wells, and its subsequent history would not have been materially different.

Chet Davis, a petroleum engineer with Richfield Oil, tested Brown in 1965–1966 by having him dowse and predict the results of wells before they were drilled. Davis chose recently permitted wells a few at a time during an eighteen-month period and asked Brown to mail him written predictions.

Davis examined the results after thirty-five wells and found that Brown had correctly predicted twenty-eight, an impressive success rate of 80 percent. Then Davis examined the seven misses—all of which Paul Brown had predicted as dry holes. Davis discovered that at each location, a dry hole had been drilled vertically, before the driller pulled back and drilled a directional branch of the well and found oil. Paul Clement Brown had been correct in all thirty-five predictions. The list of wells is not available, so we don't know many were wildcats, how many found oil. But unless the wells are all or nearly all development wells, and oil wells versus dry holes, the odds against thirty-five correct predictions would be extremely high.

Davis told writer Christopher Bird that he had met many oil dowsers in his career, and that when put to the test they all had failed—except Paul Clement Brown. Did Brown dowse for Richfield Oil? We don't know.

Paul Brown named a distinguished client list that included Getty Oil and Chevron, and he said that he had dowsed wells for Robert Minkler, president of Mobil Oil. He claimed 100 percent accuracy.[58]

"Geologists and the Geophysicists in the major oil companies are my worst enemies. All their professionalism and book-learning are threatened by my approach. I can beat them every time by a large margin when it comes to finding oil and gas."

—Paul Clement Brown[59]

Why didn't Paul Clement Brown become rich? His nephew recalled, "He was a scientist, not a businessman, which meant he didn't care about money so much." He made a good living, and kept a notebook filled with photographs of his successful wells, which he called "my babies." He typically collected 1 or 2 percent royalties on each well.[60]

Paul Clement Brown died in 1985. Shortly after, his niece and nephew were at his house, and the niece noticed her uncle's dowsing apparatus in the shed behind the house; but the next day, it was gone. Paul Brown had never worried about having his instrument stolen, because "[a]nother electrical engineer wouldn't be able to operate it."[61]

Brown told author Christopher Bird of a gold vein he divined, twenty to twenty-two feet wide, that runs from Culberson County, Texas, northward through New Mexico, to Trinidad, Colorado, a length of more than three hundred miles. The vein remains undiscovered, but it is so vaguely described as to be nearly impossible to find.

Brown also dowsed a large platinum deposit in the mountains northwest of San Bernardino, California, with twenty to twenty-two feet wide veins of the platinum mineral laurite, in a zone more than six hundred feet wide and four miles long. According to Brown, it's within the US Geological Survey Devore 7.5-minute topographic quadrangle map. Finding the supposed platinum deposit would take some prospecting—it is still undiscovered.[62]

Reprinted on page 189 of Christopher Bird's book *The Divining Hand* is a map of oil and gas deposits Brown dowsed in West Texas. Some of the areas marked on his map lie within known oil and gas fields, but many do not and still await discovery by the fortune-seeker. The undiscovered fields include some that would be hard to miss: a 500-square-mile oil and gas field just west of Marfa, Texas, and a 300-square-mile oil field a few miles east of Sanderson, Texas. The oil fields are mapped, and they wait for someone adventurous enough to drill them.

EARL PYLE AND 90 PERCENT SUCCESS

For nearly forty years Earl Pyle dowsed and drilled oil and gas wells in Kentucky and Tennessee, and apparently made a good living. At different times he claimed success rates of from 80 percent (for exploratory wells) to "better than 90 percent."[63] However, the documented results show much lower success rates.

Pyle was an Indiana farmer who started dowsing for oil in his forties. He had long been a water-dowser, but in 1952, after watching an oil dowser at work, Pyle gave his farm to one of his sons, bought a drilling rig, and started dowsing and drilling for oil near Byrdstown, Tennessee.

Pyle specialized in shallow wells, often less than five hundred feet, and sometimes less than a hundred feet deep. But along with shallow wells came smaller production rates. The big discovery of Pyle's career was a well in Lincoln County, Kentucky, that started out pumping seventy-five barrels of oil per day in 1958. In most oil-producing areas, seventy-five barrels per day would not be bragging material, but in Lincoln County it was counted a great success.[64]

Sometimes Earl worked with his brother Willie Pyle, also an oil dowser (he also claimed a success rate of 90 percent), as were Earl's wife Sarah and his son John. They experimented with different instruments, but Earl commonly dowsed with a bronze dowsing rod mounted to swivel on a handle; a white cloth dipped in crude oil was tied to the far end of the rod. His wife detected oil by a warm feeling in her palms. Brother Willie dowsed with a hollow pendulum containing a sample of crude oil.[65]

In the mid-1960s, Pyle bought a farm near Scottsville, Kentucky, sold his drill rig and quit the oil business. A few years later, Pyle and his wife bought a farm in Panama, and stayed there for a year and a half.

The Pyle family returned to Scottsville, Kentucky, and Earl Pyle reentered the oil business. Besides drilling under his own name, he formed Clementsville Oil and Gas in 1986 to drill around Clementsville, Kentucky.[66]

Geologist John W. Jewell saw the Pyle brothers in 1973 when they visited a well in northeast Tennessee just drilled by his employer, Dixie Oil. Jewell recognized them from their pictures in local newspapers but didn't know which was Earl and which was Willie. One dowsed with a V-shaped television antenna commonly called "rabbit ears," and the other had an eight-foot metal rod with a one-foot cylinder at the far end. The dowser with rabbit ears said that the well would be an oil well, the one with the long rod predicted a dry hole. They were both wrong—it was a gas well.[67]

Pyle was proud of his 90 percent success rate, but in fact, his success rate was much lower.[68] During the period 1954–1990, documents show 153 wells drilled by Pyle in Tennessee and Kentucky, with reported results (oil, gas, or dry hole). Of those 153 wells, eighty-four (55 percent) were completed as oil or gas producers, far below Pyle's claimed 90 percent, and in line with average success rates in Kentucky and Tennessee during the same time period. Statistics do not demonstrate that Earl Pyle had any oil-dowsing ability. His success is explained by widespread shallow oil in the areas where he drilled, and his low threshold for success. Some of his completed oil wells counted as successes started off at only two barrels of oil per day. Although he titled his autobiography *How to Make a Million Dowsing and Drilling for Oil*, he seems to have never become rich himself. He specialized in shallow wells that could be drilled cheaply and didn't need much oil to be profitable. He may have made his million dollars, but that amount was spread throughout a quarter century of dowsing and drilling.

Earl Pyle died in an automobile accident in 1991, at age eighty-three.[69]

WALTER J. NELSON

Walter J. Nelson was an oil driller and dowser who instantly became one of the most famous dowsers in the United States in December 1963, when the popular *Saturday Evening Post* magazine gave him a two-page write up with photographs.[70]

Nelson farmed north of Salina, Kansas, until a string of bad luck, including drought, floods, grasshoppers, and broken combines forced him out of farming in 1961. At age forty-five, he turned to oil drilling, using his dowsing rod to select drill sites. It was not the usual rod: it was quarter-inch copper tubing bent into a complex shape and filled with a mystery liquid. Nelson would grasp an end of the tube in each hand with the bent middle portion pointed forward and up at a forty-five-degree angle, walk over the land, and where the apparatus swung down, that's where he would drill. He said that an old-timer named Dewey Slick taught him to dowse for oil, but that Slick lacked the "power" to be a very good dowser. Nelson had the dowsing power, inherited from his mother.

He had tried oil-dowsing once before. In 1955, Nelson led some investors in drilling a test that turned into a dry hole. It would be six years before he tried again.[71]

Twenty people he knew had enough faith in Nelson's oil dowsing to put up the cash to drill where his bent copper tube indicated oil. The well started drilling in December 1961, and after some drilling problems, struck oil. Success attracted more investment, and Nelson kept drilling wells. By the time the

Saturday Evening Post interviewed him, he proudly told the reporter that he had drilled nine times and made all nine into oil wells.

Nelson drilled a total of thirty-one wells, thirty of them in Kansas, and reported twenty-one (68 percent) as oil producers. This would be a good rate of success, except for the small volumes of oil produced. Only about seven of the wells appear to have been profitable, a success rate of 23 percent, and not enough to make up for all the dry holes and unprofitable oil wells. The drilling program must have lost money.

Nelson's last known oil-dowsing foray was about 1971, when he dowsed a location in Iowa for the Blazer Corporation in Montgomery County, Iowa. It was a dry hole.[72]

Nelson's money-losing ways, although obscured by the fact that it was other people's money, showed in his own finances. He stopped drilling wells, fell behind in paying taxes on his oil properties, and his creditors won six court judgements, forcing him to pay more than $31,000. He moved to Wichita, where he died in 1998.[73] Walter J. Nelson, while presumably sincere, was a dowsing disaster.

THEY CALLED HIM "MR. DOODLEBUG"

Some dowsers agree with the ideas of oil geologists, only adding oil-dowsing as a way to find the oil. At the other end of the spectrum sat Clarence Elihue Hollett, whose ideas of oil were very different.

Hollett dealt in recycled rubber in Plymouth, Indiana, until he discovered oil dowsing in the 1960s. He dedicated progressively more time searching for oil with a Y-rod made of two lengths of fiberglass fishing pole tied together at one end. At the end was a sack with a secret substance that reacted with oil and gas. He said that he got the secret recipe from an old diviner who had used it to find a treasure chest in a sunken galleon off the Florida coast.[74]

Hollett ridiculed the geologists' idea that oil occurs in pools; he said that to get an oil well, you had to hit the vein, which might be only a few inches wide. To Hollett, dry holes didn't mean that there was no oil in the area; it only meant that they had missed the veins. He said that there was no forty-acre tract in northern Indiana that could not produce oil, if they would hire him to find the vein.[75]

Hollett told anyone who listened that the supposed oil shortage was a sham created by the big oil companies. In 1980, he showed a reporter a pair of large

"X" marks on a road outside Cocoa Beach and said that they had been put there by the oil companies to mark the edges of a river of underground oil flowing down the East Coast. Hollett had identified at least a trillion barrels of undiscovered oil beneath Florida. He said that once the oil companies finish mapping the oil, they planned to do away with free enterprise and impose a world government.[76]

In 1973, Hollett told a reporter that he found eight oil wells by dowsing—but all eight were ruined by "incompetent drilling." He claimed a success rate of 100 percent, which does not seem impossible if you can blame all your dry holes on the driller.[77]

Hollett advertised in Indiana newspapers for farmers to fund oil drilling on their own properties. In return for choosing the drill site, Hollett would receive a one-eighth royalty. That he found few takers he blamed on the big oil companies for telling people that Hollett was crazy. "Even my family wouldn't believe me. Nobody believed me."[78]

Clarence E. Hollett passed away in 1991 in Tavernier, Florida.[79] His family brought him back to Plymouth, Indiana, for burial. His gravestone reads:

Clarence E. Hollett Jr.

1920–1991

Mr Doodlebug

EVEN MORE OIL DOWSERS
John Moosman

John Moosman was a Swiss immigrant who farmed at St. Marys, near Parkersburg, West Virginia. He occasionally dowsed for water until an oil boom in 1889 in nearby Belmont, West Virginia, drew him into the oil business. Oil drillers rejected his advice until they noticed that his predictions were usually accurate. After he secured prominent local oil man Wilson Hurley as a client, Moosman had all the oil-locating work he wanted. In 1896, he chose a location for the Lubeck Oil Company at Cairo, West Virginia, that turned into a gusher.

Moosman entered the oil business himself, using a variety of oil-finding instruments that he carried in a satchel. One of them was a hollow pendulum filled with an unidentified "mineral substance" that caused the pendulum to swing over oil reservoirs.[80]

Fig. 13. Ohio farmer Jacob Long was known as the "oil wizard" during the Western Ohio oil boom. (*Mangum Star* (Mangum, OK), February 17, 1898.)

Moosman was deaf and didn't hear the whistle of the train that hit him in June 1897 as he walked down the track west of Parkersburg, West Virginia. He died within a few minutes. After eight years of oil-smelling, his estimated worth was half a million dollars.[81]

Jacob Long the "Oil Wizard"

Jacob Long the oil wizard was a familiar figure in the Indiana and Ohio oil fields in the late 1800s and early 1900s. Long was a farmer born in Ohio and raised in Indiana. He could not read or write and spoke awkward English, preferring the German he learned from his immigrant parents. He water dowsed for years, but one day about 1877, he noticed his forked rod behaving strangely, and concluded that it was caused by underground oil. He was said to have a strong following among oil drillers in the late 1870s. Long achieved his highest prominence in 1897, when some correct predictions put him in high demand by oil drillers. His reputation slumped after a string of wrong calls in 1898, but the lifelong bachelor remained a regular visitor to the editor of the Hartford City, Indiana, *Telegram* to offer his predictions of success or failure for drilling wells. Long died in 1914, at the age of seventy-five.[82]

Jacob R. Illyes

Professor Jacob Illyes travelled Indiana and Western Ohio in 1892 trying to convince people that he could detect natural gas with his leather-covered bottle, the contents of which were secret. A newspaper editor commented that "he looks like a tramp of the first water." He travelled in baggage cars and asked for free lodging in town jails. Illyes was persuasive despite his appearance, and many wells were drilled on his say-so. One source credited Illyes for the locations of more than a hundred gas wells in Hamilton County, Indiana, alone.[83]

Starting in 1897, Illyes bounced between insane asylums and poor farms in central Indiana. When out on his own, he tried to convince people that his pendulum told him where there were rich deposits of gold.[84] Jacob Illyes died in 1907, aged seventy.

Professor W. Taylor Hawkins

W. T. Hawkins was a Kentuckian who relocated to Arkansas City, Kansas, in 1875, and worked as a water well digger, often choosing locations with a forked stick. When oil and gas drilling came to south-central Kansas, he became so obsessed with oil and gas dowsing that his wife feared for his sanity. Then one night he heard a voice telling him how to solve the problem. After three nights with the same dream, he followed the instructions. The answer was a vial with secret chemicals that he attached to his dowsing rod.[85]

Hawkins started oil-and-gas dowsing in 1902, and local businessmen hired him to locate drill sites for shallow gas wells. He became known as a "gas witch." He was right often enough to gain some loyal clients, and he was wrong often enough to discourage others. He actively oil dowsed through at least 1916. He died in 1929, at age seventy-nine.[86]

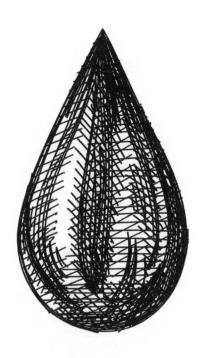

A WORLD OF OIL DOWSERS

CANADIAN OIL DOWSERS

The Canadian oil industry dates from 1858, a year before Edwin Drake found oil in Pennsylvania, and anticipating what would happen in the US, oil dowsers appeared soon after the oil discovery at Oil Springs, Ontario. Dowsing was particularly common in wildcat areas.[1] The proximity of the early American and Canadian oil fields, and the ease with which dowsers crossed the border, resulted in similar oil-dowsing practices in both countries.

The Ontario Department of Mines observed in 1928 that none of the Ontario oil fields had been discovered by dowsing. The only oil wells located by dowsing were within known fields.[2]

John H. Ethridge

John Ethridge was a Calgary oil promoter with an oil-finding instrument of some sort. His biggest success was in convincing some investors to trust his oil finder enough to drill between two dry holes in Montana's Kevin-Sunburst oil field. The result was the Ethridge-Steele #1 gusher, which sprayed oil over the landscape in April 1928. He followed up with the highly productive Ethridge-Steele #2 and the Bethel-Adams gas well. Ethridge continued dowsing for oil in Alberta and Ontario at least through 1950, but he never made another splash as large as his 1928 discoveries in Montana.[3]

Reuben Hayes

Reuben C. Hayes was a farmer from Morris County, Kansas, who went to Canada in the late 1940s to dowse for oil. Newspapers credited him for locating the well that discovered the Baron oil field in Alberta in November 1950. Geologists, however, gave credit to gravity, seismic, and subsurface geology. Hayes became a celebrity, recognizable as the only Black man in Baron, Alberta, and his oil-dowsing took him all over Alberta and Saskatchewan.[4]

Hayes dowsed with a forked stick of willow or fence wire, or sometimes a length of garden hose. He searched for oil, water, and gold, and claimed to have found the Lost Lemon gold mine for a client in the early 1950s. He was never paid more than wages, however, and his dowsing was only a sideline, as he supported himself working at a gravel pit. Reuben Hayes died in Edmonton, Alberta, in 1980 at the age of eighty-seven.[5]

EUROPEAN (AND CUBAN) OIL DOWSING

Oil dowsing was a common practice in Europe during the early twentieth century. Unlike in North America, dowsing won considerable support from European academics.

Countess Von Tüköry's Big Toe

Countess Charlotte von Tüköry's admirers thought that she had more oil-finding ability in her big toe than that of all the geologists combined— literally. When Von Tüköry walked over an oil field, she said that her big toe could feel the oil beneath. The wife of a wealthy Hungarian landowner, she was intrigued when she saw a dowser at work. She tried the forked twig herself and found that it worked for her as well. She became a socialite dowser.

Tüköry reportedly discovered an extension of the Wietze oil field in north Germany in 1911. During World War I, an oil operator brought Tüköry to the Carpathian Mountains in what is now southern Poland to find new oil fields. Tüköry was so well known that she had to travel under an assumed name, and only went out in the field late at night.[6]

Twentieth Century France: La Belle Époque of Dowsing

In the early twentieth century, French dowsers replaced the old term for their art, *rhabdomantie* (rod divination), with the more scientific-sounding *radiesthésie* (radiation detection). Ethnologist Anne Jaeger-Nosal wrote that this word change ushered in La Belle Époque (beautiful era) of French dowsing that continues today.[7]

As in the United States, French oil drilling boomed after the First World War. And just as in the United States, the 1920s in France saw a great increase in oil-dowsing. All over France, dowsers with local reputations for water-finding persuaded businessmen to bet their money on the dowser's ability to also find oil. Oil drilling attracted some of the most prominent French dowsers, including Henri Mager and Alexis Mermet.

In Europe, dowsing persisted as a common way to look for oil longer than in the US. In the France of 1925, dowsers chose most drilling locations. Numerous wells were drilled in France's Aquitaine Basin during the 1920s, often with help of dowsers, but only rarely using geologists.[8] The eastern Pyrenees area of southern France was particularly crowded with oil dowsers. The dowsers in northwest France must also have been convincing, because they not only got investors to drill a number of their dowsed locations underlain at shallow depths by igneous and metamorphic basement rock, but also to continue to drill after they hit the granite or schist.[9]

French oil drillers brought their dowsing rods to the French colonies. In Tunisia, out of twelve oil tests drilled from 1909 through 1927, two were drilled based on dowsing. Neither dowsed well had even a show of oil or gas.

Louis Barrabé, professor of geology at the University of Paris, estimated that between the two world wars, roughly equal sums had been spent drilling for oil and gas based on dowsing and based on geology. During that period no oil or gas field was discovered by dowsing, while geology discovered two, and in the years since, all discoveries in France have been made by geology and geophysics. A prominent dowser asserted that he located the discovery well for the Gabian oil field (1924), but in fact the spot had been chosen by Professor Barrabé.[10] Oil-dowsing persisted, and in 1946 a geologist complained of "the waste of considerable sums in absurd conditions" such as drilling locations in granitic terrain and on supposed rivers of oil.[11]

Switzerland

"... in no country has oil been sought as stupidly as at home."
—Elie Gagnebin, Professor of Geology, University of Lausanne, Switzerland

Switzerland is famous for its lack of every natural resource except beautiful scenery. For this reason, Swiss petroleum geologists have been like the Swiss mercenary soldiers of centuries ago: there was no work for them at home, but they were in high demand elsewhere.[12] But while Swiss geologists were finding oil in foreign countries, most exploration in Switzerland was directed by dowsers. As in other countries, the 1920s saw an increase in Swiss oil dowsing.

Dowser Alexis Mermet picked the location in 1918 for what became a dry hole near Semsales. A dowsed location at Tuggen started drilling in 1925, and did not stop until 1928, leaving a dry hole 1,648 meters (5,406 feet) deep.[13] In the 1920s, a number of companies obtained licenses to drill for oil in Vaud canton at locations selected by the swings of dowsing pendulums, but it is uncertain if any of the dowsed locations were drilled, until the 1929 dry hole at a spot chosen by the pendulum of dowser Florian Bourqui de Murist. Undeterred by his failure, Bourqui dowsed another drilling location, but the Great Depression intervened before the spot could be drilled.

A deep well located by dowsing was spud at Altishofen in July 1952. The well was drilled to 1,852 meters (6,076 feet) before being plugged as a dry hole in January 1953.[14]

In 1957, well-known Swiss geologist Arnold Heim lamented the poor placement of deep wells in Switzerland, for which he said dowsing was partly to blame. He said that the popularity of dowsing was a sign that Switzerland was "one of the most backward countries in Europe."[15]

Alexis Mermet's Pendulum

One of the most famous and influential dowsers of the twentieth century was the French-born Swiss priest, Abbé Alexis Mermet, pastor of the small village of Saint-Prex located outside Geneva.[16] It was Mermet who popularized the dowsing pendulum in France, as he dowsed for many things, including oil and gas.

Mermet announced in 1932 that he had located large deposits of natural gas near Paris, enough to supply the city for centuries. It is not known if Mermet's gas deposit was ever tested by a drill, but his huge gas reservoir has not been found.[17]

Mermet cited a number of oil test wells that he correctly predicted in Algeria, France, Poland, and Switzerland. He commented: "[F]or the radiesthetist knowing the radiations of petroleum, there is no possibility of error." Yet when his dowsing picked the location for oil drilling near Semsales, Switzerland, it was a dry hole.[18] Mermet consulted widely with oil drillers in Europe and in Algeria until his death in 1937.

Italian Dowsers

As a child in Venice, Augusta Del Pio Luogo felt painful electric shocks entering the soles of her feet and rising to the top of her head. In 1909 a dowser introduced her to the dowsing rod, which dipped in the same places where she felt shocks. She learned to dowse without instruments, guided by the shocks. The next year she moved to Argentina, where she dowsed for water.

Del Pio Luogo returned to Italy after World War I, and during the oil-drilling boom of the early 1920s companies hired her to find oil. She would press her left foot to the ground, feeling for a tremor that would rise through her leg until her whole left side shook. The stronger the tremor, the richer the deposit below. She measured depth to the substance by the duration of the tremor; for oil it was 5.3 meters per second of tremors. She experienced a distinctive taste that distinguished the different substances: water, oil, and minerals.[19]

Rachele Mussolini, wife of Italian dictator Benito Mussolini, took up oil dowsing in 1939 in a patriotic attempt to relieve a shortage of motor fuel. She was sure that her dowsing rod had discovered hydrocarbons at the family estate, Villa Carpena. Benito Mussolini requested that the state oil company, AGIP, drill his wife's oil prospect. After some foot-dragging, AGIP drilled a shallow hole and reported in January 1940 that the boring found some gas at a depth of ninety to a hundred meters, but the volume of ten to twelve cubic meters per day was too small to produce.[20]

Oil Dowsing in Germany

"A remarkable phenomenon is the resurgence of the divining rod, that spiritual and unsubstantial tool in the hands of genius, countless people, equipped with the presumed wonderful powers, strange knowledge and skills. The wonderful devout Middle Ages had the dowser pilloried, but the twentieth century began anew the hoax."

—Rudolf Delkeskamp, 1908[21]

A British mining engineer reported in 1876 that dowsing still had a following in France, Britain, Hungary, and the United States, but had all but disappeared in Germany. German dowsing started regaining popularity in the first decade of the twentieth century. In the economic and social chaos in Germany following World War I, belief in all manner of occult practices increased sharply, including dowsing.[22] A 1919 German petroleum textbook noted an increase in oil dowsing. Another German textbook on petroleum remarked that although the pre–World War edition of the book had not even mentioned dowsing, by the 1930 edition, dowsers and dowsing-type instruments were common in German oil regions: "Thus divination grew powerfully with the general nervous breakdown of the war and post-war years and the associated inclination to mysticism. It may be assumed that with the general fixing of the conditions, the changing fashion of dowsing will again abate."[23]

Dowsing continued to grow in popularity. The dowsing literature and the geology literature seem to describe different countries. The geologic literature describes areas in Germany liberally dotted with dowsing-inspired dry holes, before geology found the oil fields. The dowsing literature describes the opposite: geologists wasting huge sums of money on dry holes, when some dowsers had success rates close to 100 percent.

Dowsing was credited with discovery of a new, deeper horizon in the Wietze oil field in 1923. Oil dowser Wilhelm Meseck wrote that out of the twenty-nine German oil fields discovered through 1945, dowsing had discovered at least six or seven. According to Meseck, dowser L. Happich oil-dowsed for the German branch of Shell.[24]

Although most German geologists scorned dowsing, an exception was Johannes Walther, professor at the University of Halle, and president of the German Academy of Science. He believed that dowsers could locate water, oil, and other substances. Physicists and geologists studied dowsing, with an emphasis on physical explanations. In contrast to French dowsers, who increasingly embraced psychic explanations of dowsing, German dowsers rejected psychic theories, and the German National Association for Dowsers discouraged map dowsing.[25]

An example of the effectiveness of dowsers and geologists was the Reitbrook oil field near Hamburg. From 1915 to 1922, twelve locations in the Reitbrook area were drilled based on dowsing, and all twelve were dry holes (during the same period, two dry holes were drilled based on geology). Improvements in geophysics led to the discovery of the Reitbrook oil field in 1937.[26]

In 1929, Professor A. Kump decried the increasing popularity of diving rods in Germany. According to Kump, dowsers had located twenty-four oil wells in one district of northern Germany, but the only successful oil wells among them were those drilled in previously known fields. In the Celle district of northern Germany, outside of the Wietze oil field, forty wells were sunk through 1930 on the advice of dowsers but scored only one minor success. Eight wells based on dowsing were drilled in the Ovelgönne forest in northern Germany, resulting in eight dry holes.

Most industry periodicals criticized German oil dowsing; an exception was the London-based *Petroleum Times*, which was favorably impressed by oil-dowsing results in northern Germany in the mid-1920s. The magazine printed a praiseful full-page write-up of oil dowser Charlotte von Tüköry, celebrated in the headline as "the 'Wonder' Woman." In 1925, their German correspondent wrote: "The divining rod is a very useful help in discovering crude oil deposits."[27] The magazine editorialized in 1926:

> Nevertheless, predictions have been justified so often by successful discovery that it is difficult to avoid the conclusion that certain abnormal people have a sense denied to the majority, which is affected in a peculiar way by certain underground conditions, and which enables them to predict, for example, that water may be found by sinking for it in a given spot.[28]

By 1930, however, the *Petroleum Times* had grown more cynical, and blamed the poor record of the German oil industry on the fact that the major-ity of the well locations "in these enlightened days" were chosen by oil dows-ing: "The vogue of the diviner is chiefly due to his capacity for advertising himself; if all the successes claimed by diviners were true, Germany would have no need to import foreign oil – she would have a surplus available for export herself."[29]

Rafael von Uslar: Dowser to the Kaiser

Rafael Perfecto von Uslar was a district administrator (*Landrat*) who became the most famous dowser in Germany in 1906 when Kaiser Wilhelm sent him to dowse for water in Southwest Africa (now Namibia)—to the delight of dowsers and the chagrin of geologists.[30]

Von Uslar dowsed more than 800 places to drill for water, of which 148 were drilled, with a success rate of 79 percent. The Kaiser awarded von Uslar the Order of the Crown. However, an evaluation by the State Secretary of the

Colonial Office noted that von Uslar's wells had success rates similar to those of non-dowsed wells, and that many wells that he described as successes could not provide adequate water flows.[31]

Von Uslar landed in New York City in January 1913. He went up to the Adirondack region, to search for extensions of iron-ore deposits near Mineville, New York. His dowsing rod dipped over known iron ore, and his hosts were favorably impressed.

Von Uslar became the president of the newly formed North American Exploration Syndicate, headquartered in Passaic, New Jersey. He soon created a sensation when his dowsing detected radium deposits near Nutley, New Jersey, and coal near Port Jervis, New York. Neither the radium nor the coal was where von Uslar indicated, but before that was known, his company sent him to dowse for oil in western Canada.[32]

Von Uslar was dowsing for oil in Alberta, Canada, when World War I began. Britain declared war with Germany on August 4, 1914, and Canada followed the next day, but for the next few weeks, von Uslar continued to travel freely about Alberta, looking for oil. Von Uslar was briefly detained as a suspected German spy but was released and allowed to cross into the United States, from where he took the first ship back to Germany. The following March, drillers at a location selected by von Uslar struck oil.[33]

After World War II

German oil-dowsing, like French and Swiss oil-dowsing, appears to have had a role in oil exploration after World War II. Otto Prokop, a prominent critic of dowsing, complained in 1955 that dowsing had harmed the German oil industry: "For decades, the German oil industry, trusting in the erroneous information of the dowser, lost millions of marks in drilling, and could not procure the needed funding."[34]

Austria

In 1914, the Austrian government hired the surveyor and dowser Stanislaus Purchalia to dowse for oil in eastern Galicia (now part of Ukraine).[35]

Austria's area shrank drastically at the end of World War I, leaving the new Austria without oil fields. Prior to the discovery of Austria's first oil field at Zistersdorf, about twelve oil test wells were drilled in Austria, all of them located by dowsing, and all dry holes. When a well near Schleinbach began drilling in

1924, *The Petroleum World* noted: "Important geological opinion predicts a negative result"—and geological opinion was right.[36]

Credit for the discovery of Zistersdorf oil field in 1932 is contested. Geologist Karl Friedl insisted that dowsing had no role in the Zistersdorf discovery, but dowsers were emphatic that Viennese dowser Hans Falkinger chose the location of the discovery well.[37]

Emerich Herzog was born in Vienna, Austria, in 1869, and emigrated to the United States in 1889. In the New York City directory for 1892, Herzog is listed as an importer. He became a naturalized American citizen in 1892, but soon after he returned to Austria on his US passport and settled in Vienna.

He later said that he learned to dowse before he left America at age twenty-two, and that he "discovered the gushers of Oklahoma, Oakland, and Mexico."[38] That is nonsense. When he sailed back to Europe in 1892, the gushers of Mexico and Oklahoma were still years in the future, and Oakland—if he was speaking about the city in California—is still waiting for its first producing oil well.

Near the end of World War I, Herzog published two books on dowsing. He claimed wonderful power with his dowsing rod. In 1922, some California oil drillers said that they were importing Herzog to oil-dowse for them, but no results were announced. In 1926, Herzog announced that his dowsing had found a huge oil deposit in Austria—about 3.7 billion barrels.[39] Unfortunately, no oil boring has ever found this supposed oil field.

Herzog travelled to Britain in 1936 and dowsed for oil both from an airplane and on foot, finding supposed oil fields in Hampshire, Sussex, and Devon. After he spent one month in Britain, the *Guardian* mistakenly reported that five of the professor's oil locations near Brighton had been confirmed by actual production.[40] In fact, Herzog was unable to interest the oil companies in drilling any of his locations.

Emerich Herzog was Jewish. During World War II he was taken to the Theresienstadt concentration camp, where he died in June 1943 at age seventy-three.

British Oil Dowsing

There was very little oil dowsing in Britain prior to 1936 because of a lack of private oil drilling. Companies drilled numerous oil test wells from 1936 to 1939, but it is not known if any wells drilled were based on dowsing, although a number of dowsers offered their services.

A dowser from Cardiff wrote to the *Petroleum Times* that his method was to walk over the oil prospect with a bottle of Boots' Medicinal Liquid Paraffin in his pocket, which sensitized him so that he could feel oil underneath, and then define the limits of the pool. He assured the magazine's readers: "In my oil-finding work I leave nothing to chance."[41]

Mining engineer G. Percy Ashmore announced that his dowsing defined a large oil pool beneath Sussex Downs—enough, he said, to supply the nation for more than two hundred years. Canadian oil dowser William Partridge dowsed for oil around England and told the press that he had found oil in Devon.[42]

Capt. F. L. M. Boothby returned to Britain from Argentina, where he said that he had successfully dowsed oil. At the suggestion of Col. Arthur Bell of the British Society of Dowsers, Boothby and another dowser, Maj. K. W. Marylees, dowsed southeast England and found what he considered two attractive oil prospects, one in East Sussex and the other in Dorset.[43]

The government granted a license to the Midlothian Petroleum Syndicate to drill southeast of Edinburgh, Scotland. The syndicate was formed in 1936 by geologist E. H. Cunningham-Craig to drill an offset to the D'Arcy well drilled in 1919–1922, which was abandoned after producing only forty-nine barrels of oil. Cunningham-Craig thought that the D'Arcy well should have been drilled deeper—he would get his chance to test his prediction.[44]

Charles A. Pogson joined the company to work on the regulatory and financial side.[45] Pogson had served for more than four years as the official water dowser for Mumbai, India, and was credited with a success rate of greater than 95 percent.[46] Now in England, he dowsed a number of the proposed oil locations, and the Midlothian well was the only one where his dowsing instrument responded.

Anglo-American Oil drilled and completed the well for ten barrels of oil per day in July 1938—which in oil-thirsty Britain was counted as a success. Col. Arthur Bell, president of the British Society of Dowsers, wrote that both a geologist and a dowser had correctly predicted oil at the Midlothian well, and that it should be the model for cooperation between dowsing and geology. Cunningham-Craig objected, writing that he would never use a dowser.[47]

Evelyn Penrose, International Dowser
Miss Evelyn Penrose was born to dowse. She was born on Midsummer Day, which was believed to bestow occult power. She was born and grew up in

Cornwall, where belief in dowsing was strong. Her father had inherited dowsing ability from her grandmother, and he passed it to Evelyn.[48]

Photographs of Penrose do not show a beautiful woman, although she was said to have "sharp" blue eyes. Her autobiography recounts the apparent romantic event of her life, when as a young woman in a Belgian boarding school she shared a perfect—and perfectly proper—evening of ballroom dancing while being romanced by a young Belgian Army captain. Her autobiography mentions no other affection from men, but lack of romantic complication left her free to explore other adventures.

Evelyn Penrose preferred being called a "radio perceptionist" rather than dowser or diviner. She first dowsed for oil while visiting California in 1929. At the Signal Hill oil field, she discovered that walking over underground oil gave her severe nausea and headaches, and a very good oil reservoir would cause her to black out. Although Penrose found map-dowsing for oil less debilitating than on-site oil dowsing, she could only map dowse in half-hour sessions once or twice a day and was often left with blinding headaches.

The Canadian province of British Columbia hired Penrose in 1931 to find water during a drought. Her appointment created a stir, both because she was a dowser and because she was a woman. The BC government also had her dowse for oil. She discovered that she could feel oil fields up to eleven miles away using upraised palms facing in the direction of the oil but was so sickened from the effects of oil beneath her that she spent two weeks in a hospital.[49]

After Canada, Evelyn Penrose dowsed through Africa, India, the Caribbean, and South America, before settling in Perth, Western Australia. Evelyn Penrose was hired to find oil in Canada, the United States, and other countries, but not in her home of Australia.[50] She blamed the geologists' disdain for dowsers partly on dowsers' misconceptions about oil:

> At the risk of making myself unpopular with my fellow-diviners, let me say that I have a good deal of sympathy for the geologist in his contacts with the divining fraternity, for if they have to listen to the same sort of nonsense that is inflicted on me in nearly every part of the world I visit, it isn't surprising that there are so many prejudiced and antagonistic geologists.
>
> Perhaps the worst are those who want me to join them in their wonderful discoveries, such as "rivers of oil flowing across the country," which they have followed with their divining rods for at least twenty miles.[51]

The Michigan oil operator Don Rayburn asked her to map-dowse for oil. After some back-and-forth, Penrose returned one of Rayburn's maps with a 2,000-acre oil reservoir outlined in red pencil. According to a September 1954 letter from Rayburn, he drilled ten wells based on her map, and got nine oil producers; the lone dry hole was mistakenly drilled outside Penrose's outline.[52] However, the record does not support Rayburn's nine oil wells drilled into Penrose's map-dowsed outline. From February 1952 to September 1954, Don Rayburn drilled ten wells, of which eight were dry holes. It is possible that Rayburn drilled the wells under another name.

Miss Evelyn Penrose died in 1962 at the age of eighty.

Narcisso's's Rod

American Mexican wildcatter Mordelo Vincent hired Belgian Narcisso Barbiou to dowse for oil in the Tampico, Mexico, oil boom during the 1910s. Barbiou used a wire V-rod with a five-peso coin soldered to the tip. Before he could find any oil for Vincent, Barbiou was diagnosed with syphilis, and was treated by ingesting the arsenic compound arsphenamine. He suddenly lost his oil-dowsing ability and blamed the arsphenamine.[53]

Cuba

Small oil companies, both Cuban and foreign, used dowsing in the early 1950s. By the Cuban Revolution of 1960, active Cuban oil dowsers included Favier, Foyo, and Pairol, but they failed to find oil. A farmer named Urquiza invented an instrument to search for water, oil, and minerals, but it is not known if any oil tests were drilled based on his doodlebug.[54] Dowsing was reintroduced to Cuba by Soviet and Czech geologists in the 1970s, but the Cuban state oil company (CUPET) does not seem to have used dowsing to locate oil wells.

It's Not Dowsing; It's Biolocation

You wouldn't think that the former Soviet Union would be where dowsing was most accepted by mainstream scientists. Dowsing was suspect in the Soviet Union because it apparently violated Marxist materialism. Although Communist ideology rejected the occult, there have always been exceptions, such as Stalin himself consulting the psychic Wolf Messing.[55] The Soviets respected practical results, and so tried dowsing. It was in the Soviet Union, and now Russia, where dowsing was and is the most accepted by science.

Some Soviet geologists openly used dowsing for mineral exploration in the late 1960s.[56] It achieved a breakthrough in 1967 when V. S. Matveev published in a Kazakh geological journal his study showing that the dowsing rod reacted across lead and zinc deposits. Matveev used a dowsing rod made of iron, copper, or brass, and called it the "biophysical method."

Matveev and coworker N. N. Sochevanov brought the biophysical method to wider attention in 1974 when they published a paper on dowsing in the all-union journal *Geologia Rudnykh Mestorozhdenii* ("Geology of Ore Deposits"). Sochevanov and Matveev wrote that there were then about forty geological groups in the Soviet Union using dowsing. Knowing that the article would be controversial, the editors put a note on the first page of the article acknowledging that the method had critics. In the same journal the following year, Schmidt, Eremeev, and Solovkin called dowsing a "shameful superstition," and declared that it "has nothing in common with scientific methods."[57]

In 1979, Christopher Bird characterized dowsers as "an important minority" among Soviet geologists and geophysicists.[58] Although highly controversial, dowsing was taken seriously in the Soviet Union and remains so in present-day Russia, discussed in mainstream geological journals. The terminology officially changed in 1979 from the "biophysical" or "biogeophysical" method to the term used today: "biolocation."

The official government newspaper *Pravda* printed a scathing rebuttal of dowsing in 1982 and quoted a top geophysicist as saying: "Their 'successes' are pure fantasy." The article described a rigorous Soviet study of dowsers: "They confirmed yet again that dowsing is superstition."[59] But the ill opinion of *Pravda* did not stop the spread of dowsing. In 1990, V. K. Kozyanin noted in a Soviet journal: "In 1989, the Kamchatka oil and gas prospecting expedition (KNGRE) carried out much biolocational [dowsing] work which showed a high correlation with drilling data in the known fields."[60]

A 1992 article in the English-language magazine *Science in Russia*, published by the Russian Academy of Sciences, described results of choosing drill locations from satellite photos by geologists, L. Uvarov and Yu. Konnov, in which they were so successful that they were regarded as clairvoyant. In eight drilling programs in Russia, the geologists had a 76 percent success rate, roughly double the expected success rate of 30 percent to 40 percent. The author reported that their forty-five successes out of fifty-eight borings in Western Siberia had a chance probability of less than one in ten thousand.[61] Although Uvarov and

Konnov did not dowse the locations, the addition of a dowsing pendulum and the ideomotor effect would presumably achieve the same results.

Mardanov and others in 2017 proposed that oil dowsers were reacting to electromagnetic fields generated by the petroleum. They documented numerous studies that showed dowsing anomalies centered on oil fields. The method has also been discussed as a way to prospect for oil recoverable by massive hydraulic fracturing.[62]

Biolocation survived the breakup of the Soviet Union, and it is taken seriously by some geologists in Russia and some parts of the former Soviet bloc, including Bulgaria, Romania, Czechia, Ukraine, Poland, and Estonia, as well as Cuba, although in some of these countries, dowsing seems to be used mostly in other fields, such as archaeology, and not in oil or mineral exploration. Other scientists in these countries are hostile to dowsing.[63]

The scientific attention dowsing receives in the former USSR is far greater than in the West, where it is ignored by geologists and respectable geological publications.

INTRODUCTION TO DOODLEBUGS

WHAT IS A DOODLEBUG? PART II

"Doodlebug" is American slang dating to the 1860s referring to various insects, most commonly the ant lion (genus *Myrmeleontidae*). It was no doubt an extension of "doodle," meaning a foolish person, which goes back centuries in England ("Yankee doodle" was not a compliment). In early twentieth century America, "doodlebug" was applied to any foolish person or thing, especially politicians and small vehicles, such as small railroad engines. Doodlebug entered the British vocabulary during World War II as their defiantly belittling name for the V-1 rocket.[1]

Doodlebug as an oil and gas exploration method first saw public print in 1914 to describe a pseudo-geophysical device. As slang, the word was tossed around loosely, making it sometimes difficult to know what is meant. The term spread to dowsing instruments such as the forked stick and the pendulum. When genuine geophysical techniques arrived in the early 1920s they were also called doodlebugs, and the name stuck. This book uses "doodlebug" to describe pseudo-geophysical devices only.

How Does That Doodlebug Work?

Sadly, we have no idea how most historical doodlebugs worked, because the inventors kept their operation secret, and today most doodlebugs have disappeared, presumably thrown out with the trash. Following are the known mechanisms.

> Ideomotor action: The operator unconsciously causes the mechanism to react. This is what causes dowsing instruments to move, and so applies to any doodlebug such as a hand-held pendulum or swivel-rod that resembles a dowsing tool.

> Subjective sensation: The reading relies on subjective changes in touch or sound experienced by the operator. Examples are radionic devices, which depend on a subjective stickiness to the touch.

> Random or noncorrelative readings: The instrument reads a real and nonsubjective physical field, but one which is unrelated to oil deposits.

> Operator manipulation: This involves conscious fraud by the operator.

Confusing Doodlebuggers with Geophysicists

The way geophysicists toss around the term doodlebug confuses those outside the industry. Calling an instrument a doodlebug implies that it is unscientific and worthless. Applied to people, doodlebug or doodlebugger may or may not be derogatory. Seismic field crews are called doodlebuggers (or jug hustlers, "jug" being slang for a geophone), with no thought of disrespect. But the editor of the *Oil & Gas Journal* lamented that in these ultrasensitive times, "jug hustler" might be too offensive, and feared that "doodlebugger" might not be far behind.[2]

Geophysicists commonly refer to themselves as doodlebuggers, in part because the original meaning is nearly forgotten. It is common between friends who can insult one another in terms that would cause offense coming from a stranger. Also, in calling themselves doodlebuggers, geophysicists show the spit-in-the-eye-of-snobbery attitude admired in the oil business. The same ethos used to inspire many in the industry, from roughneck to company president, to casually refer to one another—among themselves—as "oil field trash." The industry's big Christmas party each year in Denver was, until a few years ago, officially named the "Oil Field Trash Bash." If you are going to speak of doodlebuggers or oil field trash, smile when you call them that.[3]

Early Oil Doodlebugs

Although doodlebugs didn't really catch on until the 1900s, pseudo-geophysical oil-finding devices go back to the early years of the oil industry. William Reed came to the Pennsylvania oil fields in 1859, dowsing for oil with a forked twig. He soon switched to a machine of his own invention to find oil. His instrument was described as a tube hanging from a tripod. He never explained it other than that it operated by electricity, but he was reported to be very successful. He believed his doodlebug could also detect gold. He drowned in 1887 while searching for sunken treasure from a skiff off the shore at Atlantic City, New Jersey. His mysterious oil- and gold-finder was lost to the sea.[4]

D. M. WATSON: EARLY OREGON DOODLEBUGGER

The Spindletop gusher of 1901 at Beaumont, Texas, was in an area previously thought by most geologists not to have significant oil potential. The discovery inspired entrepreneurs in Oregon, another area thought to have no oil, to organize six oil companies in 1901 and 1902. Soon six wells were drilling away.

D. M. Watson was a popular restauranteur in Portland who sometimes drilled water wells. When Oregon's first oil boom hit, Watson told newspaper reporters that it had taken him thirteen years to perfect a machine that infallibly revealed the presence of underground oil by detecting some sort of earth currents. He said that he had tested it in California, and that his services were in great demand in Texas and Pennsylvania.

Watson predicted that a well at Monmouth, Oregon, would hit oil at less than 1,000 feet. He visited another well drilling nearby, and again predicted oil at less than 1,000 feet. Both wells were dry holes. Newspaper reporters overlooked his failures in Oregon and printed Watson's tall tales of fabulous oil discoveries elsewhere: California, Louisiana, Pennsylvania, and Texas. In 1904, when another wave of drilling started, Oregon newspapers called Watson an "oil expert," and quoted his moonshine about the Willamette Valley being

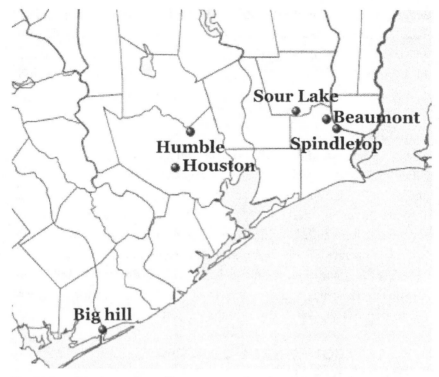

Fig. 14. Localities in southeast Texas mentioned in the text.

incredibly rich in oil. He also endorsed false accounts of prolific oil deposits at Pratum, in Marion County.[5] The wells at Pratum were abandoned as dry holes, but that did not put a dent in fawning newspaper coverage of Watson as a great oil expert. The *St. Johns Review* in Portland described him in 1909 as: "a man whose career in oil finding has been nothing less than phenomenal since he has not made a single mistake in locating oil in the past five years, nor during that time has any oil been discovered at any point where he advised against boring."[6]

News reporters didn't question how Watson could be supposedly so infallible in faraway places, yet always fail locally. Watson and his machine found an oil deposit near Oregon City in 1909, so he organized another company, and drilled another dry hole. Watson's failure at Oregon City finally ended his doodlebugging career.

HOW TO DRILL DRY HOLES AND STILL GET RICH

It was one of the most legendary land deals in the history of Texas oil. In 1913, Dr. Peyton Standifer Griffith was in hard negotiation to buy a tract called the

Stevenson Survey, north of Houston near the small town of Humble. It was 1,400 acres of pine and scrub oak, but the saloonkeeper, a man named Kahler, was determined not to sell cheaply. The nearby Humble oil field, where wells pumped oil from the shallow caprock of a salt dome, was nearly exhausted. There was talk that more oil might be found by drilling deeper, off the edges of the Humble salt dome, though to that point it was mostly just talk.

As the popular version of the story goes:

> He bought the fourteen-hundred-acre Stevenson Survey from a Mr. Kahler, who had a saloon on the main street of Houston. Dr. Griffith offered to buy the property by spreading twenty thousand dollars in twenty-dollar bills on the bar. Mr. Kahler accepted the offer and told Dr. Griffith he was a damn fool.[7]

It's a great story, and although it isn't true, well, it ought to be.

The story of the transaction changed through the years for greater theatric effect, but the result was even more dramatic than the embellished details of the sale. Griffith and his partner, the oil driller R. E. Burt, bought one of the most oil-soaked parcels in Texas, the Stevenson tract. Griffith had seemingly overpaid for the unproven land because he knew without doubt that there was oil there, and lots of it. His doodlebug told him so—and his doodlebug was right.

Griffith said that he invented the doodlebug himself, and how it worked was nobody's business. Many doodlebuggers called themselves "doctor," but Griffith had a genuine medical degree. He had left medicine for a business career, in such varied pursuits as undertaker, running a feed mill, and making shovels. He became prosperous but had not yet found his true calling. Then in 1902, when he was thirty-nine, he read about the Spindletop oil boom. He had dabbled in oil drilling in Tennessee and saw money to be made. He caught a train to Beaumont, Texas, bought an interest in an oil well on his first day, and made money from the start.[8]

Dr. Griffith's oil-finding machine, a doodlebug before the oil industry started calling them doodlebugs, was a box with three flexible tubes sticking out. Two short tubes served as handles, while the longer third tube went into Griffith's mouth. A coil stuck out of the front of the box. To the end of the coil, Griffith would attach a metal capsule the width of a pencil and about three inches long. He had different capsules to detect oil, gas, salt water, sulfur, silver, or gold. Although he never revealed what was in the capsules, he said that his machine operated on the principle of affinity: like affects like, so the capsules presumably contained samples of the substances being searched for. When Griffith walked above the sought-for substance, the capsule would dip down.

Ever the gentleman, Dr. Griffith always strove for a dignified appearance, even when carrying his oil-finding machine with the tube stuck in his mouth. He had a neatly trimmed beard, and always wore a stovepipe or derby hat and a Prince Albert coat. He would precisely outline the oil field and moved drill rigs as little as six or eight feet, to be inside the field limits as defined by his machine. It was easy to make fun of the doctor and his odd contraption—it was less easy to mock his success.

Dr. P. S. Griffith was one of the small number of doodlebuggers who became rich; even rarer, he became rich by investing his own money. And yet his doodlebug led him to drill mostly dry holes. How was this possible?

Griffith and his doodlebug first came to public notice when he discovered oil in the Big Hill Salt Dome, in Matagorda County, Texas. Big Hill isn't very big, but on the flat coastal plain it qualifies as Big Hill. Like Spindletop Hill, it is the surface expression of a salt dome. Others had drilled at Big Hill and found only gas, which was nearly worthless for lack of a market.

But one well had seen a good show of oil before it was plugged. The discouraged owners sold the property to Dr. P. S. Griffith, who chose a location guided by his doodlebug. In May 1904, the Griffith well blew out at a depth of 850 feet, spouting 2,500 barrels of oil per day. The gusher created a sensation, and operators rushed in to drill more wells. During the next two years, drillers put down thirty-four more wells, of which at least fourteen produced oil. Griffith himself drilled ten wells. The oil at Big Hill went quickly to water, and production died by the end of 1905. The oil producers sold only 198,000 barrels of oil from Big Hill—an additional unknown quantity of oil soaked the land and ran down the creeks during blowouts.[9] Yet it made a name for P. S. Griffith and his doodlebug.

In 1904, the year that he struck oil on Big Hill, Griffith moved his family from Tennessee to Houston. If his doodlebug could find small oil fields, it could find big oil fields, and Houston was a growing city with opportunities for an astute businessman such as P. S. Griffith.

Griffith was a good oil man, but most of all, he was an excellent businessman. He put all his money and effort into Texas oil and Houston real estate. They would both be good long-term bets. Griffith invested in more oil wells at the Batson and Sour Lake fields. He did well, but his great success was still ahead.

Griffith believed that the oil along the Gulf Coast had leaked up from the Trenton Formation, which he believed was 5,500 to 6,500 feet deep beneath Houston. The Trenton is an oil-bearing formation in the Appalachian Basin, but is not even present, as far as anyone knows, beneath the Gulf Coast. It was precisely this wrong theory of Gulf Coast oil that led him to success.

Griffith was not the first one led to success by a wrong oil theory. Anthony F. Lucas believed that both salt domes and their associated petroleum were the result of volcanic activity. Both suppositions were wrong: neither salt domes nor petroleum are created by volcanoes. But his wrong theory gave Lucas the confidence to drill the Spindletop gusher.[10] Although Griffith was wrong as to the specific source of the oil, he was partly right in that petroleum is generated deep in the ground. Griffith's deep Trenton theory, as well as his oil-finding machine, gave him the confidence to drill ever deeper—where he struck oil.

In 1913, oil drillers were leaving the Texas Gulf Coast, because the Gulf Coast was running out of oil. The Spindletop gusher started an oil boom, but it led oil drillers down a blind alley. Oil men drilled the caprock of shallow salt domes, looking for another Spindletop, but discovered that few shallow salt domes had oil in the caprock. They discovered the Sour Lake and West Columbia fields in 1902, Batson in 1903, and Humble in 1904, but then the discoveries stopped. The caprock oil fields were quickly drilled and exhausted. Oil production on the Texas Gulf Coast went from nearly nothing in 1900 to twenty-eight million barrels in 1905, then crashed back down to less than six million barrels in 1913.[11] No major fields had been discovered since Humble in 1904. They had drilled all the obvious salt domes, and many surface oil seeps. The Gulf Coast looked like a boom that had busted, an exhausted oil province of no lasting importance, and drillers left the coast for the big fields being found in North Texas and Oklahoma.

There was a lot more oil on the Texas Gulf Coast, but drillers didn't know where to look. They had done some tentative drilling around the edges of the salt domes and found some oil, but not enough to get excited about.

Then in 1913, drillers stepped off the edge of the Sour Lake Salt Dome, and hit good oil reservoirs in the deeper sands around the flanks of the dome.[12] This discovery turned around the fortunes of the Gulf Coast oil industry, and Griffith and his doodlebug were ready to ride the new boom.

Humble Oil Field: How to Drill Dry Holes and Get Rich

Griffith must have seen that the deep discovery at Sour Lake was a game-changer. His doodlebug confirmed that there was also oil around Humble. Griffith and his partner, veteran oil driller R. E. Burt (who later became mayor of Dallas), bet their fortunes that Griffith's doodlebug was right.

According to oil industry lore, P. S. Griffith bought the 1,400-acre Stevenson Survey, from a saloonkeeper named Kahler, by spreading a thousand twenty-dollar bills on the bar. Well, it makes a good story. The reality is that in June 1913, Griffith and Burt bought the Stevenson Survey, 1,394 acres, from Friedrich E. Koehler. Koehler, the son of German immigrants, had started out running a grocery store (not a saloon), but by 1913 he was a major Houston real estate developer. It is doubtful that an experienced businessman would have been swayed by such theatrics as flashing a stack of cash. Koehler sold the Stevenson Survey for $15,000 down and another $40,760 paid during the next few years. Griffith and Burt paid top price for the Stevenson Survey from one of Houston's most astute businessmen. The only way they could make money would be if Griffith's doodlebug were right, and the Stevenson Survey held much more oil than anyone had a right to expect.[13]

Koehler had leased portions of the land for oil drilling, but no one had found much on it. R. E. Burt had himself drilled on Koehler's land. In 1906, he and another partner had drilled down to 2,000 feet on the Stevenson Survey, and even found oil, but it didn't last long. Now Griffith's doodlebug convinced him that the big oil was deeper down, and that it would be worthwhile to buy the Stevenson Survey. As expensive as it seemed in June 1913, buying the Stevenson Survey would be by far the best deal Griffith ever made, for within a year it became the most valuable oil property in Texas.

Oddly, Burt and Griffith seem to have found no oil at Humble. The prolific deep oil sands at Humble were discovered by the Producers' Oil Company in October 1913, when the #10 Carrol started off producing 1,500 barrels of oil per day. Then a few weeks later, Producers' Oil #11 Carrol roared in at more than 10,000 barrels per day, before settling in at a steady 6,000 barrels per day from 2,700 feet.[14]

Meanwhile, guided by Griffith's doodlebug, Burt and Griffith drilled their #1 Hamblen to 3,250 feet; it was a dry hole. They drilled more wells, penetrating successively deeper. Even while other oil companies were hitting gusher wells around Humble, Griffith and Burt, using their own money, drilled a succession of deep dry holes at Humble. Yet they still became rich from Humble.

Although Griffith and Burt didn't drill any producing oil wells at Humble, they were wise enough to see that it mattered less who found the oil than who owned the mineral rights. Griffith believed the philosophy later expressed by J. Paul Getty: "The meek shall inherit the Earth—but not its mineral rights."[15] By aggressively leasing or buying large tracts, Burt and Griffith came to control 20,000 acres (about thirty square miles) surrounding Humble Dome. They made sure that they would own a large portion of any oil found near Humble, no matter who discovered it.

Even while they were drilling dry holes, Burt and Griffith leased part of their land to other companies and profited from the royalties when their rivals hit headline-grabbing gushers on Burt and Griffith leases. Griffith even used his oil finder to find two drill locations for Humble Oil Company (Humble instead drilled halfway between the two spots, a location Griffith said would be dry, but they hit oil anyway). Eventually, Burt and Griffith gave up the risky business of drilling at Humble and leased all their land there to other companies. The great surge of oil from the wells around Humble made Burt and Griffith rich. The largest oil company at Humble was the Producers' Oil Company, later bought by the Texas Company, better known as Texaco. Texaco had Griffith confirm their early drilling locations at Humble.[16]

Did Griffith's doodlebug find oil at Humble? He thought that it did, and that gave him and his partner the confidence to snap up leases and land around Humble Dome. Griffith's doodlebug failed him in choosing specific locations to drill, but he became rich on his royalties.

Drilling the deep sands around the flanks of the salt domes stopped the downward spiral of the oil industry on the Gulf Coast. Texas Gulf Coast oil production rose until it exceeded the 1905 peak of the Spindletop boom and kept rising. Griffith's doodlebug helped convinced him to get out ahead of the industry and made him wealthy.

Humble secured both Griffith's fortune and his reputation as an oil finder. Griffith and his oil finder—some called it an electric machine, some said it was magnetic—were in great demand. He travelled to Mexico and to numerous US states, but he was wrong far more often than he was right.

Griffith carefully surveyed the Balcones Fault through Limestone County, and pronounced it dry, although Mexia and other rich oil fields were later found there. While vacationing in Colorado, Griffith doodlebugged a location north-east of Colorado Springs in 1919, and drilled there on and off for the next three years until he finally gave up in 1922. Griffith drilled other dry holes in Texas, Indiana, and Illinois. He told reporters that he had detected a large oil field a few miles south of Oakland, California; if true, the oil is still undiscovered.[17]

According to family lore, while searching for oil in Ohio, his companions jokingly asked him if his doodlebug could direct them back to town, thinking that their circuitous route had confused Griffith. But Dr. Griffith consulted his machine, which responded when he faced in the proper direction toward the town. He explained to his surprised audience that he had plugged in his gold-finding cartridge, which pointed the way to the bank vaults in town. Late in his career, Griffith took his doodlebug to Mexico to search for buried trea-sure. He tuned his instrument to silver and noted the locations along his railroad route where his machine reacted.[18]

One man who watched Griffith on numerous occasions concluded that the machine was just for show, and that Griffith chose locations based on surface clues. Another switched the metal capsule that bobbed up and down over oil and saw that the machine worked just the same with a capsule meant to detect something other than petroleum.

According to historian Mody Boatright, after Humble Field, Griffith's oil finder lost him more money than it made.[19] But Griffith was an astute busi-nessman, and more than offset oil-drilling losses by strategic investments in Houston real estate. The many dry holes drilled on the advice of Griffith's doodlebug suggest that it was worthless as an oil finder. The great value of his machine was that it gave Dr. Griffith the confidence to buy the Stevenson tract.

Dr. P. S. Griffith died in 1937, a wealthy and respected Houstonian. What happened to his doodlebug is not known.

WILBUR MCCLEARY AND THE FIRST DOODLEBUG

Interest in oil-finding instruments picked up in the early twentieth century, and the dribble turned into a flood. When a thing becomes common, it demands a name. "[I]n fact oil can be located in any place by this 'doodle bug' as it is popularly known here."[20]

In July 1914, the Wichita, Kansas, *Beacon* printed an article about Wilbur McCleary, an undertaker from Altus, Oklahoma, who was prospecting for

oil in Kansas using an oil-finding machine of his own invention called the
"doodle bug." This is the earliest known reference to an oil-finding machine
called a doodlebug.[21] Other things had been called doodlebugs before—insects,
small railroad engines, automobiles, and a popular dance—but McCleary's is
apparently the first oil-finding machine to be so called. All future doodlebugs
would be named after McCleary's original.

McCleary was born in Iowa and moved his family to Oklahoma about
1907, where he worked as an undertaker, first at Oklahoma City, then in Altus,
Oklahoma. But he spent his spare time tinkering with a treasure-locating gizmo.
McCleary's instrument was an elaboration of the traditional V-shaped dowsing rod. It was two sticks joined at one end, while the other ends were held
one in each hand, palm up. McCleary carried a dry-cell battery in his pocket,
connected to wires that ran one each up the prongs of his rod. Where the rods
joined, the wires connected to a secret mechanism wrapped in a piece of oil
cloth. He had actually meant to invent a gold-finder, but when he noticed that
it reacted to a can of kerosene, he adapted it to locating oil.[22]

McCleary tried his machine at the Healdton field in Oklahoma, and the
Burkburnett oil field in Texas. He could use it while walking the ground, or
riding in a train or automobile. People tried fooling him. They hid or buried
bottles of crude oil, they blindfolded him, but the doodlebug always dipped
over the oil.

After McCleary detected oil a few miles northeast of Altus, businessmen
formed the Jackson County Oil and Gas Company, with McCleary as president,
to lease 5,000 acres guided by McCleary's doodlebug.[23]

McCleary's fame spread, and oil operators brought him to Kansas, but
the doodlebug stumbled. While being driven around the Augusta oil field,
McCleary declared one well a dry hole, but it was one of the best oil wells in
the field. After a moment's embarrassment, and some rechecking on foot, he
came up with three excuses: his machine was overwhelmed by the escaping gas
around the field, he had been sitting too close to the automobile's gasoline tank,
and the well had actually missed an exceptionally rich oil vein by only forty
feet. McCleary also visited a well being drilled near Wichita and predicted a
dry hole; but a few days later the well came roaring in as a large gas producer.

Back at Altus, the Jackson County Oil and Gas well began drilling in
November 1914, and reached a depth of 1,200 feet by March 1915, but ran out
of money and had to dig deeper into their pockets. After a two-week hiatus, the

well resumed boring downward, but no oil or gas was found, and the company plugged the dry hole.[24]

The businessmen at Altus were disillusioned, but Wilbur McCleary still believed in his doodlebug. He sold his undertaking business and became a full-time oil locator. He chartered the Wilbur Oil Company in 1916, and later the Elk-Trail Oil Company. He lived at Rockwall, Texas, then Elk City, Oklahoma, then Tulsa. His level of success is unknown. He stayed in the oil business, but never again made headlines as he had in 1914.

Wilbur McCleary died in 1921, in Oklahoma City. By that time the *Oil & Gas Journal* was referring casually to the "doodlebug method" of finding oil. McCleary had lived to see the term "doodlebug" adopted industry wide.[25]

WHEN DOODLEBUGS RULED THE EARTH

"The doodlebug is abroad in the land; the woods are full of oil smellers; men claiming to be endowed with some supernatural second sight that enables them to locate oil and gas by walking over the ground, are found everywhere."
—Charles N. Gould, 1922[1]

THE EARLY 1900S WAS DOMINATED BY THE ACCELERATING PACE of science, and many tried to harness the new science—as best they understood it—to oil-finding.[2] Pseudo-geophysical instruments popped up here and there, but they were far less common than traditional oil dowsers until the decades 1900–1920, when the dribble of doodlebugs turned into a flood.

The classic doodlebug was a machine hidden in a box, known in the oil fields as the black box. It was usually small enough to be carried by the inventor as he walked around searching for an oil signal. For practical purposes it could not be too large to sit on the seat of a car, as the inventor twirled the dials and the car drove down the road on its search for oil.

How doodlebugs operated is usually lost to history. Even when the inventors explained its operation, listeners were usually no better informed. The instruments usually operated by some garble of radio/magnetism/

Fig. 15. A satirical view of doodlebugs. (*Natural Gas*, September 1922.)

radiation/vibrations, but little that made sense to a scientist. The doodlebugger usually knew the vocabulary of science better than the principles of science.

It might have been doubtful that black-box doodlebugs worked better than a forked twig, but what did science have that was better? Oil companies increasingly hired geologists during the 1910s, but effective geophysical tools did not arrive in the US until 1922.

No one was sure what worked or not, and trial and error was the order of the day. Oklahoma oil scout L. E. Sears noticed that his automobile stalled when over oil fields and decided that the engine had become dilapidated to the point where it was affected by electromagnetic emanations from oil reservoirs. The next time his car stalled, he leased the land and convinced his employer to drill. The wildcat well hit oil and produced greater than 550 barrels per day.[3]

The number of oil dowsers did not decline with the rise of doodlebugs. Oil dowsers also increased from 1900 to 1920, but not nearly as much as doodlebugs. From 1920 to 1950, doodlebugs in the United States seem to have outnumbered oil dowsers. An observer of the Southern California oil boom in 1920 noted that the boom created business for both dowsers and doodlebuggers.[4] Dowsers and doodlebuggers were more allies than opponents. They both saw that their largest obstacle was the skepticism of the geologists and geophysicists. Many dowsing devices became complicated to the point that it was difficult to draw a line between dowser and doodlebugger.

Tom Edison and the Self-Taught Inventor
In 1877, an unknown Thomas Edison walked into the New York office of *The Scientific American* and placed in front of its astonished editors a

working phonograph, a machine that talked! Edison was an American original: self-taught and both ignorant of and disdainful of theory.[5] He instantly became the patron saint of every lone tinkerer with modest background and big dreams.

After Edison, no one could safely sneer at unknown geniuses, unstifled by formal education and unburdened by scientific theory. Americans idolized the practical, unschooled inventors of the Henry Ford mold. Schoolchildren learned about Edison inventing the light bulb through exhaustive trial and error, of Charles Goodyear inventing vulcanized rubber in his kitchen, and about two bicycle mechanics who built a flying machine. These were American icons.

The enduring popularity of the Tom Edison type is shown by that of his young fictional counterpart, Tom Swift. A self-taught boy inventor with no education beyond high school, Swift has starred in more than a hundred boys' novels, starting with *Tom Swift and His Motor Cycle* (1910), and including *Tom Swift and His Great Oil Gusher* (1924). The fact that there was never a title *Tom Swift and his Electric Doodle-Bug* seems just an oversight. Early twentieth century scientists dazzled the public with a bewildering variety of new and mysterious forces: alpha rays, beta rays, gamma rays, n-rays, x-rays. Certainly, one of these could be harnessed to find oil. What would Edison do? Enterprising tinkerers set to work.

The Radio Age

Radio was the sensation of the age. In the late teens and early 1920s, daily life was transformed by radio voice transmissions. Radio waves spread silently, invisibly, instantly—magically—through miles of air, right through your solid brick walls and into your living room. Out of radio sets in isolated farmhouses came the sounds of singers and orchestras playing big-city ballrooms. Build-it-yourself radio kits were popular, giving hands-on training in theory and construction of radio receivers and transmitters. Many doodlebugs supposedly detected oil by radio waves. There was the radioscope, radio oil finder, radio cameraphone, radio seismograph, radio emanator, and the radiologeometer.

Doodlebugs came tumbling out of basements, garages, and barns too rapidly to keep track. It seemed in the 1920s that every town in Kansas had a farmer who invented a doodlebug in his tool shed. Each new oil discovery in Texas and Oklahoma attracted a swarm of doodlebugs, each genius eager to prove that he had invented the infallible oil-finding machine.

Most doodlebugs failed, as quickly and unmourned as bugs under a streetlamp. With luck, the editor of the small-town weekly might give them a brief mention on page two, in the column of short paragraphs on doings around town. Most of them tried and failed and went back to the farm. Yet a few doodlebuggers made long careers out of their inventions.

Doodlebugs were most common in oil-rich areas where the risks were moderate. Doodlebugs appeared wherever wells were being drilled, but they seemed particularly common in California and in the mid-continent region, which included north Texas, north Louisiana, Arkansas, Oklahoma, and Kansas. In 1920, a geologist estimated that there were at least 150 dowsers and doodlebug operators active in the mid-continent. The head of geophysical research for Humble Oil, Dr. Ludwig Blau, kept an open-door policy in his Houston office, and between 1930 and 1936 he examined more than 1,000 oil-finding devices brought in by their inventors.[6]

DOCTOR ABRAMS AND RADIONICS DOODLEBUGS

The first chapter of this book introduced a black box in Calgary, the electrical device without a power source. That was a radionics, sometimes called psionics, device. The boxes are called Hieronymus machines, Wigglesworth Pathoclasts, or Rifle machines, after various inventors, but they all spring from the imagination of a San Francisco physician, Dr. Albert Abrams.

Albert Abrams was an MD from the University of Heidelberg who built electrical apparatuses he called a reflexophone and a sphygmobiometer, used to diagnose disease.[7] The patient stood with each foot on a metal plate, holding an electrode to his forehead. The operator adjusted the dials on the reflexophone until the settings harmonized with the vibrations of the patient's condition. The doctor would notice a slight increase in stickiness of a rubber or glass rod rubbed across the patient's abdomen. Abrams didn't need to examine the patient: he could substitute a drop of the patient's blood, a strand of hair, a signature, or a photograph.

In 1919, Abrams introduced the Oscilloclast, which he said would emit radio waves to cancel the radiations given off by diseases in the human body, including tuberculosis, cancer, and syphilis. With each treatment, the Oscilloclast reduced the "ohmage" of the disease until the disease was completely extinguished.

Abrams rocketed to national fame when the novelist—and fan of fringe medicine—Upton Sinclair wrote a fawning article in *Pearson's Magazine*.[8] Abrams taught his methods to others by charging $250 to observe him at work for four weeks. His disciples could then lease his equipment for an initial fee of $200 to $250, plus five dollars per month, but they had to agree never to open the cases. The agreements were not always honored, and knock-off versions of Abrams's machines appeared. Traditional doctors saw their practices suffer.[9]

In 1916, Abrams wrote that everything, including water, petroleum, and minerals, broadcasts electromagnetic waves. He claimed that he could tune his sphygmobiometer in on the desired substance, and even determine its depth and quantity. In February 1923, he announced that he had perfected an oil-locating machine, which he would soon exhibit, but it appears that he never did.[10]

Scientific American magazine appointed a committee to examine the Abrams method. The magazine printed a twelve-part series, in what the British journal *Nature* described as "the characteristic American style," evidently referring to the blunt criticism. The committee found that the Abrams method failed in both theory and practice.[11]

The *Scientific American* committee took apart Abrams's instruments and described "a collection of crude electrical apparatus." Not only did Abrams's machines fail to correctly diagnose but the equipment itself was superfluous. During a demonstration of an Abrams box, one of the investigators, unnoticed, disconnected a wire, and saw the operator continue to obtain measurements uninterrupted. The committee derided Abrams's published works as "weird and rambling" and "an incoherent hodge-podge." Abrams confused fundamental electrical concepts of frequency, current, potential, and resistance. It was nonsense in theory and useless in practice, but they decided that Abrams was self-deluded rather than a fraud.[12]

Abrams died in 1924, leaving behind an estimated 3,500 medical practitioners using his system, and additional unknown thousands of imitators. From the 1920s through the 1960s, at least fifty Abrams imitators built and marketed their own electronic medical-treatment gizmos.[13] The machines made their way into oil exploration as ready-made doodlebugs.

George de la Warr, an English civil engineer, started making radionics machines in the 1940s, and incorporated his own designs, including a camera. By 1951, he claimed to be able to photograph scenes from the past.[14] Writer Allen Spraggett described the inside of a de la Warr black box as: "a jungle of wires with a rubber band and a bar magnet thrown in." He also noted that there was "no power source, no battery or plug-in."[15]

De la Warr used radionics machines in oil exploration. In 1953 the Burmah Oil Company paid de la Warr £13,000 to use his radionic camera in Oxford to create photographs of places in Burma (modern Myanmar) where they could find oil. The experiment was a failure.[16]

O. W. KILLAM: DOODLEBUGGER WHO STRUCK IT RICH

"Oh, I haven't used any very scientific method. It's kind of a little geology and a little doodle-bugology and a little common sense and a few things put together. And in reality, I've done most of my own thinking along that line rather than taking other people's advice."

—O. W. Killam[17]

The common saying in the oil patch is that doodlebuggers don't get rich. You claim that your machine finds oil, so why aren't you rich with your own oil company?

Unlike other doodlebuggers, O. W. Killam became wealthy. He gave some of the credit to the doodlebug he carried around, literally a "little black box" whose contents he kept secret. Doodlebuggers have hit successful wells now and then, and in rare cases have discovered substantial fields. But successful doodlebug wells are almost always in established oil territory; Killam discovered oil in a county where oil had not previously been discovered.

Killam built his own oil company from scratch, using his doodlebug. What was inside it? Some people believed that Killam's black box contained—of all things—a bottle of urine. Joe Morris, one of Killam's drillers, thought that the box was empty.[18] Killam wasn't saying, except that it worked by chemical reaction.

He was christened Oliver Winfield Killam, but everyone knew him as O. W. He grew up in a small town in Missouri, and from an early age, his goal was to become a millionaire, like the captains of industry he read about in the St. Louis newspapers. Taught by his mother that he could achieve anything he wanted, as long as he worked hard and didn't quit, he never doubted that he would become rich—it was only a question of how.

He became a lawyer. He later modestly said that he was no good at lawyering, but he appears to have succeeded at everything he did. He moved to Joplin,

Missouri, then to Oklahoma. For a while he worked as a laborer in a lumber yard, but then became a lawyer and businessman in Missouri and Oklahoma and served multiple terms in the Oklahoma state senate. He promoted lead and zinc mines in Arkansas and southwest Missouri. Mineral dowsers were numerous around the lead and zinc mines, and Killam probably saw doodlebugs at work there.

In 1920, O. W. Killam was forty-five years old, married with three children, owned a thriving business, and had a promising political career when he did something wholly unexpected of a successful, middle-aged family man: he quit his home and career, and started over in unfamiliar territory in an unfamiliar business. Killam sold his business interests, gave up his political career, and moved his family to Laredo, Texas, where he bought a drilling rig and started drilling for oil. Killam was a stranger in South Texas, and he summed up his oil experience as: "I didn't know anything about the oil business."[19] At that time, South Texas seemed a poor place to find oil. Drillers had found some gas and hints of oil, but South Texas was unglamorous poor pickings compared to Oklahoma, where gushers were being found regularly.

Killam started drilling where few people thought there was oil. He chose an area in Zapata County because the topography reminded him of the oil-rich area around Bartlesville, Oklahoma. O. W. drilled a couple of dry holes, but in April 1921 the beginner in the oil business, the newcomer to South Texas, completed the Hinnant #3 well in Zapata County, the discovery well for the Laredo Field, and the start of the prolific Mirando Trend of oil fields. It was the first commercial oil well in South Texas.[20]

O. W. Killam continued to find oil with the help of his doodlebug. When he wasn't using his doodlebug, Killam would sometimes drive around South Texas, occasionally stepping out of the car to fire his shotgun. He would listen for an echo, and when he heard what he wanted, he put a wooden stake in the ground, and drilled at the spot.

In 1956, Killam gave an interview to the Oral History Project at the University of Texas. Everyone knew about Killam's doodlebug, but Killam dodged questions about it. He said that there were any number of doodlebugs, and it didn't matter which you used. He said that the ground surface or vegetation above oil fields always seemed to have a different appearance, implying that a successful doodlebugger reacts to subtle surface clues. He said that, although doodlebugs were not very accurate, neither was geology, so he tried to use a bit of everything.

Killam described a doodlebug as: "A little instrument that goes up and down and around and around and makes you spend your money drilling holes in the ground, and the law of averages will finally hit you a pool of oil."[21] But he also believed that petroleum emits a sort of radiation that would affect certain chemicals, and that if he could discover the right chemicals, a doodlebug could become a reliable direct-detection device for oil. In the meantime, he warned that doodlebugs were not very accurate and needed to be combined with common sense.

Killam said that his most successful geologist was a one-armed, one-quarter American Indian who did his best work surveying the terrain by climbing up telegraph poles. Perhaps, though others recalled that when visitors came to the rigs, Killam would sometimes send a roughneck to the top of the drill rig to pretend to be his famous one-armed American Indian, by tucking one arm out of sight inside his shirt and making a show of scanning the countryside for the next drilling location.

O. W. Killam seems to have been successful in doodlebugging not because his doodlebug was infallible, but because he knew that no oil-finding method is infallible. He knew not to trust any single method, and to keep his wits about him. His formula was "a little geology and a little doodle-bugology and a little common sense and a few things put together."[22]

Killam told his interviewer that he wouldn't recommend the doodlebug to anyone else, but it had done well for him: "And I've had a lot of fun with it and I've had a lot of outdoor exercise and I've kept up my enthusiasm and good health. And I've made enough to buy the groceries. And what more do you want?"[23]

O. W. Killam could be self-effacing to a friendly interviewer, but he would also stand up to criticism of his methods. At a geology conference in Corpus Christi, when a geologist sarcastically asked him about his doodlebug Killam shot back: "I have a million dollars. How much do you have?"[24]

Killam died in 1959. Today, Killam Oil is still an active family-run enterprise in South Texas. The little black box has disappeared, replaced by modern technology. What was in that black box? O. W. Killam wouldn't say.

MEXIA: BOOMTOWN FOLLIES

It all came together: the spiritualists, seers, dowsers, doodlebuggers, and geologists, in that free-for-all of drilling and gushers known as the Mexia boom. Col. A. E. Humphreys drilled the discovery well at the little town

Fig. 16. A cartoonist's view of the Mexia boom. (*Oil Weekly*, October 8, 1921.)

of Mexia, Texas, in 1920 on the recommendation of self-taught geologist F. Julius Fohs. But geology was an uncertain tool, and drillers, including Humphreys, hedged their bets with methods that claimed more certainty, such as psychics.[25]

The pattern of Texas oil boomtowns had been set by Beaumont following the Spindletop gusher: the twenty-four-hour crowds, the hustlers, fortune-tellers and fortune-seekers. Geologists joined the throngs in the nineteen-teens. Then after World War I, the doodlebuggers showed up in force at every oil boomtown.

It seemed like a who's who of doodlebuggers assembled, drawn to the spotlight that briefly shone on Mexia.

A writer for the *National Oil Journal* assured readers that Mexia had many seers, dowsers, and doodlebuggers. Like the twenty-four-hour commotion, the cries of newsboys and hot-dog vendors, these were the signs that told him that this was a real oil boomtown, like Ranger, Breckenridge, and Burkburnett.[26]

Seers . . .

The most prominent clairvoyant in town was Mexia native Annie Webb West, whose psychic oil-finding advice was rewarded by Col. Humphreys with an $8,500 gift.

W. F. Wittich, "psychic oil geologist," moved to Mexia in July 1921.[27]

Another psychic claimed that six months previously he had suddenly and mysteriously acquired the power to know when he was above an oil reservoir. He tried to convince oil operators of his superpower by predicting the fate of drilling wells, with mixed results.

Mexia also had its share of clairvoyants who may have given oil drilling advice, but did not mention oil-locating powers in their advertisements. These included Madame Lily, Madame C. Le Honda, and Madame Merselle Cleo.

Dowsers . . .

Dowsers were there. One had a "turnip-shaped" object on a string, evidently a Mermet-style pendulum.

And Doodlebugs

Doodlebugs were rapidly increasing in popularity, and it was in doodlebugs that Mexia excelled. The Fort Worth *Star-Telegram* reported that there were more doodlebugs in Mexia than in any other oil boomtown, each out to prove that his own invention was the true oil finder.[28]

T. D. Hall of Electra, Texas, was there with his Chemigraph, a versatile instrument capable of detecting buried oil, gold, and silver. It was described as a small instrument made up of copper wires connecting vials of chemicals.[29]

Dr. G. W. Cook of Houston showed up with a radio magnetic instrument, which he claimed had already located successful wells.[30]

M. C. Trumbull came over from Louisiana. Although he had sold twenty-five-year licenses for the use of his affinity instrument to nine other companies, he considered Mexia important enough to go there himself.[31]

Abner Davis showed up, of course, as he seemed to have a nose for places where there was foolish money running loose. Mexia was such a place, where whatever good judgement investors had was washed away by visions of gushers dancing in their heads. Davis had failed repeatedly in banking and oil but was now touting San Antonio inventor M. S. Blackburn's radioscope as a foolproof oil-finding device. Blackburn used his invention to locate another structure east of town, supposedly a look-alike to the Mexia structure—but it proved a dry hole.[32]

Fort Worth oil man J. E. Pope came to town to drill a location selected by the inventor of a Petroleum Magnet. He declared that he had passed up on a gusher and a fortune during the oil boom at Burkburnett, Texas, by not listening to the Petroleum Magnet's inventor, but that he was now ready to risk his money drilling where the Petroleum Magnet indicated.[33]

GENERAL ROBERT A. LEE

Most oil doodlebuggers are sincere in their belief that they can find oil. The same cannot be said of company promoters. Robert Aaron Lee was out to prove that he could find oil, but the organizers of his namesake General Lee oil companies didn't care if Robert found oil; the company organizers made money by selling stock, not by finding oil.

Robert A. Lee made the rounds of oil companies in Fort Worth but found no one who would give his oil-finding machine a chance until he met Charles Sherwin and H. H. Schwartz, who were looking for a way to attract investors. They built a company around dowser Lee, named the General Lee Oil Company.

Sherwin and Schwartz fabricated a family connection between Robert *A.* Lee and Confederate Gen. Robert *E.* Lee, and in 1922, newspaper ads across the United States named Robert A. Lee the "miracle man of the oil industry." His picture, with a white beard like his supposed ancestor, featured prominently. The organizers promised enormous profits: investments would pay off six-to-one within thirty days. The pitch was so successful that they quickly followed up with the General Lee No. 2, and then the General Lee Interests. The sales campaign raised more than $195,000, but most of the share sales went through a Chicago brokerage that absorbed a 37.5 percent commission. The companies kept no account ledgers, so what happened to the investor money was a mystery—but not so much a mystery that anyone doubted in whose pockets it landed.[34]

Robert A. Lee was born in Tennessee in 1847, enlisted at age sixteen in the *Union* Army, where he served for a year and a half. He moved his family to Texas after the war. Lee moved to Spokane, Washington, where he worked as a stone mason, then lived in Boise, Idaho. When he met Sherwin and Schwartz, he had been back in Texas for three years, and was working as a janitor at the county courthouse in Denton, Texas.[35]

Robert Lee was unhappy with the false publicity, but he wanted to prove that his invention could find oil, so he put up with the lies. But in August 1922 he rebelled when Sherwin and Schwartz drilled only two wells, both dry holes, and were uninterested in drilling more. In promotional letters under his signature, he had promised shareholders that at least ten wells would be drilled. He demanded an accounting, but Sherwin and Schwartz refused. Lee sued to make them drill more wells, and in December they agreed to set aside $25,000 to drill at least eight more.[36]

The General Lee Oil Company was caught in a legal crackdown on fraudulent oil-company promotion in Fort Worth. In April 1923, a federal grand jury indicted ninety-two Fort Worth oil company promoters for mail fraud, including such luminaries as polar explorer Frederick Cook, and flamboyant promoter and aviator S. E. J. ("Alphabet") Cox.[37]

Sherwin, Schwartz, and Lee went on trial in May 1923. The prosecution brought in relatives of Robert E. Lee to testify that they were unrelated to Robert A. Lee. Shareholder after shareholder told the jury how they were fooled by the company literature.

Robert A. Lee on the stand was an "old, bent man." He testified that he really could find oil. He drew his doodlebug out of his pocket. It was a hybrid doodlebug/dowsing rod: a small box containing copper wire and a secret mixture of chemicals, with two holes in the box to insert the ends of wooden rods, forming a traditional V-shaped dowsing rod. Lee called it a chemical battery. He didn't know how it worked and didn't know what chemicals it held. It was given to him about 1911 by a widow in California whose husband had made a fortune in oil.[38]

The prosecutor recommended that Robert Lee be let off with a warning, but the judge sentenced him to two years and a $6,000 fine. Robert Lee entered

Leavenworth federal prison in June 1923 and was released on parole after eight months.[39]

Charles Sherwin and Henry H. Schwartz appealed, and while out on bond they went to Mexico to run another fraud, the Mexerica Mining Company, before the US Postal Service cut off service. Sherwin and Schwartz joined the Florida real estate boom until their final appeal was denied, and they entered the federal prison at Leavenworth. In 1932, Charles Sherwin went into the television business: not in the business of making or selling televisions, of course, but into the business of selling shares in the Ray-O-Television Manufacturing Corporation, which he misrepresented as having a factory on Long Island, New York, where sets were rolling off the assembly line. A federal jury convicted Charles Sherwin of mail fraud in 1939, and the court sentenced him to five years in prison.[40]

Robert A. Lee moved back to Spokane, Washington, and lived quietly until his death in 1934, at age eighty-seven.

REVEREND OLSON'S DOODLEBUGS

In 1922, David Olson was driving his family cross-country from Minneapolis to San Francisco, by way of Oregon. He had previously been a small-town minister in Oregon, and his old friends saw that he was still tall, handsome, and charismatic, but now he drove a car that few preachers could afford: a twelve-cylinder Packard, the leading American luxury car of the era. The vehicle was not just any Packard. What attracted press coverage, first in the Bismarck, North Dakota, *Tribune* when he drove through that town, and then in the Portland *Oregonian*, was that David Olson's Packard had a radio! In 1922, a car with a radio seemed futuristic, out of a Jules Verne novel. David Olson had become rich in the oil business, he said, because he had invented instruments that could detect oil deep in the ground with 100 percent accuracy.

Was the Rev. Dr. David Eugene Olson a credulous believer in his oil-finding machines, or a shameless liar? He split his career between spreading the gospel and drilling for oil guided by worthless oil-finding machines. Born in 1877, the son of immigrant farmers in Iowa, David Olson grew up speaking Swedish. He enrolled at the Eugene Divinity School in Oregon in 1900. After an interruption to serve as minister in Cottage Grove, Oregon, he received his diploma in 1908.

Olson was an impressive man, more than six feet tall, with broad shoulders, blue eyes, and golden-red hair. His speech was as striking as his appearance, and his unstoppable energy, nimble mind, and absolute self-confidence gave him charisma that held audiences spellbound. In 1913, he founded the Minnesota Bible College in Minneapolis, and became its president. Olson had undeniable preaching ability and commitment to his faith but was a poor business manager and did not work well in an organization. He resigned from the college in 1919 and moved to California, which was in the midst of an oil boom. Olson joined his brother Carl and the oil promoter W. A. Sage to drill wells and learned how to doodlebug for oil.

Olson returned to Oregon in 1923 and announced that he was one of the foremost oil geologists in the nation and that he had visited nearly all the oil and gas fields in the United States, Canada, and Mexico. Olson had an assortment of oil-finding machines, which he had tested at thousands of wells in thirteen states—he said they never failed. How he accomplished all this travel, study, and testing in such a short period is a mystery, but he spoke with a forceful conviction that banished doubt.

Oil had not been found in Oregon, although about thirty test wells had been drilled. Olson's machines determined that Oregon had more oil than Texas or California and he chose two drilling locations over large oil fields: one two miles south of Eugene, on a rise called City Outlook, owned by Eugene Divinity School; and the other outside of Cottage Grove, on top of Mt. David. Olson was well-known and well-liked in both Eugene and Cottage Grove.

Olson followed a pattern familiar to con men. He announced that there was no stock for sale, because the project was entirely funded by savvy California oil people. Of course, he conceded, a few shares could be made available to friends in Eugene, as a personal favor. He soon relented, and said that it would be a shame if the people of Eugene were unable to share in the profits; he would sell shares—but only to people in Oregon.[41]

Olson organized the Guaranty Oil Company of Oregon. The company issued 500,000 shares of stock, but gave 300,000 of those shares to David Olson, brother Carl Olson, and their California partner W. A. Sage, in return for 6,000 acres of oil leases, probably bought for a dollar an acre or less, and some drilling equipment. The organizers donated 100,000 shares to the Eugene Divinity School. Drilling money was provided by selling the remaining 200,000 shares for a dollar per share. The corporation was no longer content to sell only to Olson's selected friends, or to those asking to buy shares, but sent out salesmen to hard-sell to the doubtful.

Olson exhibited his four different doodlebugs, which he said had taken him many years to perfect, at a cost of hundreds of thousands of dollars. One doodlebug was curved in a gooseneck and pointed the direction to any oil deposits within a mile. Another, with the appearance of an ordinary forked dowsing rod, would detect oil beneath the spot. The other two doodlebugs checked the results of the first two.[42]

David Olson gave lectures at Eugene and Cottage Grove on oil drilling mixed with religious sermons and hymns. To those who wavered, Olson offered the clincher: he was so sure that his doodlebug locations at Eugene and Cottage Grove, Oregon, would find great oil fields that he personally guaranteed to refund all the investors' money if oil were not found. He gave his word as a follower of Jesus that those who bought shares in his wells stood no risk of losing even a penny.

What more could anyone ask?

Olson's religious colleagues and his former congregation knew him as a devout man, and knew he must be sincere. Olson recruited other ministers to serve as officers and board members of the company. He received a glowing endorsement from President Eugene C. Sanderson of the Eugene Divinity School. Olson's money-back guarantee completely won over the Eugene *Morning Register*, which, under the headline "Lucky Lane County," editorialized: "What a fortunate community this is! Elsewhere people must struggle and strive, take risks and deny themselves. Here they may become rich without even taking a chance."[43]

With a money-back guarantee, and endorsed by the president of the local seminary and a local newspaper, why not get rich? People mortgaged their houses to buy shares, and Sunday school teachers sold shares to their classes.

Not all were favorably impressed. The editor of the *Capital Journal* in Salem accused Olson of using religion to sell worthless shares in his oil company, and went after him in editorials titled "An oily preacher," "Oily Olson gushes," and "Doing the doodle-bug."[44]

Eugene was also home to the University of Oregon, whose geology professors warned the public away from Olson's moonshine. To geologists, the Willamette Valley was an unlikely place for oil. A number of test wells had been drilled in the valley, and geologists had seen nothing encouraging. Olson's oil-finding machines, they recognized, were pure hokum, and they said so.[45]

Olson turned the controversy into an underdog-against-the-establishment story, and portrayed himself as an oppressed Galileo. He said that the professors

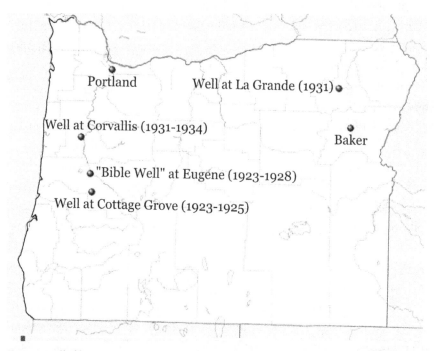

Fig. 17. Drilled locations in Oregon, all dry holes, selected by Rev. David Eugene Olson's doodlebugs.

were either ignorant or had an "ulterior motive." Olson wrote that "big interests" wanted the oil fields for themselves. He called geologists "foolologists," and sneered that the geology professors believed that men descended from apes, and so why should anyone believe them about oil? He said that the reason geologists did not use instruments such as his was that the doodlebugs would put the geologists out of work.[46]

The university geologists were evidently uncomfortable with public controversy, but Olson gloried in controversy. In the end, the scientists had only the facts on their side, which proved no match for Olson's persuasive oratory and lies. David Olson was selling dreams of easy wealth, and people wanted to believe him. But the scientists at the university did their public duty, for which they were insulted and abused by David Olson, with no discernable effect in slowing down sales of Guaranty Oil stock.

The Olson brothers and their partner W. A. Sage repeatedly guaranteed that if the wells were dry holes, they would refund the investors. They wrote that their combined assets were worth much more than a million dollars, so that they could easily refund the money—not that there was the slightest chance of failure. The Olson brothers certainly looked like rich men, driving about in Packards. Backed by Olson's money-back promise, the investments were absolutely foolproof. Guaranteed.[47]

Drillers began the well on land owned by the Eugene Divinity School in June 1923, and in July, at a depth of only 475 feet, the company announced that oil had been struck, and that five to ten barrels of oil per day were being bailed from the well. This was probably just black mud, but it created excitement. Olson said that his instruments showed the main oil pay at 1,300 feet, but the drill bit passed 1,300 feet without any oil, and kept drilling.

The well near Cottage Grove began drilling in September 1923. By January 1924, the well depth was 400 feet.

Olson invited two prominent California doodlebuggers to vouch for his location. C. L. Cofer and his Terrestrial Wave Detector came to Eugene, where he found three oil sands below the Eugene well, ready to gush out millions of barrels of oil. Samuel Spitz used his petrolometer to determine that oil sands were present at depths of 3,100 and 4,500 feet beneath the Eugene well site, and at a depth of 3,630 feet at Cottage Grove. All three instruments, those of Cofer, Spitz, and Olson, were regarded as worthless by the oil industry. Olson later used, and claimed to have invented, the "Oilometer," which appears to be a knock-off of Spitz's petrolometer.[48]

On April 19, 1925, at nearly 3,000 feet deep, the well owned by the Eugene seminary blew out. Water and mud shot out of the hole, higher than the 104-foot-tall derrick, followed by gas, estimated by the Guaranty Company as a million cubic feet per day. People from Eugene drove out to the well site, causing such a jam that police were dispatched to direct traffic.[49]

A chemist at Oregon State tested the gas and found that it would not burn because it was just air that had been trapped underground, not natural gas associated with oil deposits.[50] The newsmen and their readers didn't catch the significance of the gas analysis, however, and Olson announced that the university was once more twisting facts to slander him. The excitement over the gas blowout was Olson's opportunity to sell more stock, and none too soon, because the $160,000 initial capital was nearly exhausted. At a shareholders' meeting, Olson painted an exciting picture of wells at both Eugene and Cottage Grove about to penetrate—possibly any day now—the great oil deposits that were certainly just beneath the gas. Outside interests, he warned, wanted to snatch the oil bonanzas away from the people of Oregon. He again offered his money-back guarantee, this time with 8 percent interest, to anyone who wanted to sell their shares.

Drilling continued, but as the weeks, then months, went by, it was clear that an oil bonanza was not imminent, and there was possibly a long way yet to drill. In April 1925, Olson mailed letters to shareholders asking them to send voluntary contributions of ten dollars per month to keep the rigs drilling.

In December 1925, the company stopped drilling the Cottage Grove well, which was down about 2,000 feet, to save scarce cash for the Eugene well. Olson wrote to shareholders again in October 1926, asking for contributions of two dollars per month, but by December 1, 1926, the money again ran out, and the company stopped drilling the Eugene well at a depth of 3,050 feet. Olson proposed a 5 percent assessment on all shares, except those held in escrow. This meant that everyone would have to pay the assessment except David Olson, his brother, and Sage.[51] The assessment was blocked by President Sanderson of the Eugene Divinity School because it would be a great burden on the college.

David Eugene Olson left for Montana in 1927 and let others try to salvage something from the financial wreckage. By 1928, David Olson was predicting (by mail) that the Eugene well would find oil sands at 3,300 feet and 4,400 feet. He (falsely) denied that he had ever predicted oil at shallower depths.[52]

Neither well found oil.

A Eugene woman asked the Guaranty Oil Company for the promised refund of her $500 investment, but the company insisted that it could not refund money. In September 1925, she sued the Guaranty Oil Company, W. A. Sage, and David Olson, asking the court to enforce their money-back guarantee. Olson blamed the lawsuit on his enemies, but others demanded refunds, and more lawsuits followed.[53]

Despite Olson's repeated guarantee, he never refunded any money. An investor convinced the court to order three town lots owned by Olson in Cottage Grove to be auctioned off at a sheriff's sale. After that, investors collected from the rapidly shrinking funds of the Guaranty Oil Company. Latecomers found themselves suing a corporation without assets.

The growing suspicion that David Olson's money-back guarantees were worthless was confirmed when a Eugene couple sued in March 1929 to enforce a similar guarantee for money-back-plus-interest on their $10,000 investment in two of Olson's wells in California.[54]

By December 1926, the Guaranty Oil Company had spent all its capital, and was in debt to the driller, W. E. Marrion, who reluctantly filed a lien against the company. The only asset the company had left was the oil lease on the land beneath the drilling site, so the lease went up for auction at a sheriff's sale, and the driller bid $4,844, the amount of his lien. With a few others who had come to own the drilling equipment via similar liens, Marrion organized the Crown Oil Company and began drilling at City Outlook once more.

Crown Oil drilled until it ran out of money. The company stopped paying its drill crew in March 1928, and by July, the hole was 3,600 feet deep, past the depth of 3,300 feet where Olson's doodlebug indicated oil. The crew filed a lien for $2,400 on the well. Drilling halted again, this time for good.[55]

When the drill crew foreclosed on the well and planned to pull out the well casing for salvage value, shareholders in both Guaranty and Crown cried foul. Pulling the casing to sell it is standard practice at a dry hole, but without casing, the hole would collapse, and shareholders would lose their last, slim hope. But the court sided with the drill crew and the casing was pulled.[56]

The corporation commissioner revealed that investors had lost about $250,000. Many had mortgaged their homes, lured by the heavenly oratory and empty promises of David Olson. In Montana, David Olson told a reporter that if his Montana wells came in, he might return to Eugene, reenter the old hole and drill deeper to find the oil he knew was there.

What little hope that remained ended in the early morning of May 25, 1929, when a fire of unknown origin consumed the wooden drilling derrick. The Eugene *Register*, perhaps embarrassed by its previous enthusiasm for Olson's oil scheme, was glad that the derrick was no longer visible on the promontory south of town. The newspaper recalled: "High-powered salesmanship was resorted to in selling the stock. Religious faith was played upon. . . . Into a highly speculative enterprise thus were drawn the dollars of the poor."[57]

The oil derrick atop Mt. David at Cottage Grove blew over during a windstorm in 1933.

David Olson wasn't around to watch the collapse of the dreams he inspired in Oregon. By that time, he was living in Minneapolis, and doodlebugging for oil in Montana.

Olson took his doodlebugs to Montana in the midst of an oil boom. He was still a compelling speaker, and his lectures on geology in Montana were well-attended. Olson called himself a "scientific petroleometer geologist," and with his usual persuasiveness, he convinced many that his oil finder could not fail. Companies hired Olson to doodlebug their locations and featured his endorsement in newspaper ads selling stock. Olson's doodlebugging earned him seats on the boards of directors of startup oil companies.

Olson was using a stick grasped at one end and held horizontal. Over an oil field, the free end of the stick bobbed up and down, indicating a depth of one hundred feet for each bob. Olson had many clients in Montana, but there were also oil industry people who recognized Olson's method as a fake, and the *Montana Oil Journal* derided his doodlebug, calling it "a late 1928 model with everything but a cowcatcher."[58] He replied that his instruments were infallible.

In April 1928, Olson's endorsement of a drilling location was featured in large newspaper advertisements selling shares of the Kalispell-Niarada Oil Company. It was a dry hole, but Olson had an uncanny knack for finding new clients even after repeated failures.

Olson's wildcat location in Chouteau County for the Geraldine Oil Company was a total bust. The well went down 970 feet without finding oil. Olson insisted that the well had drilled through an oil sand at 600 feet. He was so persuasive that the company skidded the drill rig over a few feet and drilled a second well, to test Olson's mystery sand. The second well passed the 600-foot depth without any oil and didn't stop until 2,170 feet—it was another dry hole. Other failures included the Kenny #1 and the Kalispell-Polson #1. Olson chose two locations for the Hibbing-Sunburst Oil Company in the Kevin-Sunburst field. The wells found oil, but picking locations in the middle of the oil field was not much of a risk. Olson ended 1928 with six dry holes at wildcat locations, and two oil wells drilled inside a known field.

Olson returned to Montana in 1929 as head of the International Royalty Holding Company of Canada. He drilled a pair of wildcat dry holes, then quit Montana for good. In 1931, David Olson joined in drilling two dry holes in northern Washington state.[59]

Businessmen in La Grande, in northeast Oregon, asked Olson to prospect near their community. In exchange for $2,000, Olson detected large oil deposits southwest of the town, stacked one atop the other at depths of 1,000, 1,600, and 2,600 feet. The townspeople started drilling in May, but they had drilling problems, and had reached a depth of only 390 feet in August before they ran out of money and confidence. Despite a pep talk by Olson, the businessmen reckoned they had spent enough.

Olson's wells at Eugene and Cottage Grove had failed, bringing a flood of bad press and hard feelings after people lost money they couldn't afford to lose based on Olson's lies. Yet he was so charismatic that despite all the failures, he talked investors from Portland into drilling south of Corvallis, Oregon. The well started drilling in early 1931 and went down in fits and starts until it was abandoned in 1934. It was Rev. Dr. David Olson's last dry hole.

Olson quit the oil business in 1932. He accepted the pastorship at the struggling Alvarado Church of Christ on Sunset Boulevard in Los Angeles. Listeners started packing into the church to hear Olson's sermons, and membership skyrocketed. Olson built up the youth program and established a church summer camp and a sixty-voice choir. His unstoppable energy suddenly stopped in 1940 when he suffered a stroke. He struggled to resume his preaching, but died from a second stroke in April 1943.

According to his family, Olson died a pauper. He owned no home, had no bank account, and no pension or life insurance to provide for his ailing widow. Tributes poured in from those touched by his powerful preaching. Hardly anyone remembered his doodlebugger career.

Was David Eugene Olson a conscienceless liar or a deluded visionary with a dowser's unlimited but mistaken faith in his instruments? Perhaps Olson was both sincere *and* a crook. It appears that he truly believed that his machines could find oil. He lied shamelessly, but perhaps he believed that his lies were all for a good cause. He may have promised to repay his Oregon investors only because he believed that there was no chance he would have to pay.

For all the money he misspent on his dry-hole doodlebug locations, he apparently didn't keep any for himself. He probably spent his own money on dry holes as foolishly as he wasted other people's.

SAMUEL SPITZ, INVENTOR AND DOODLEBUGGER

Samuel Spitz was a prolific inventor, and if his fantastic inventions didn't always work, they always made great news stories. His oil doodlebug, the petrolometer, was ineffective in finding oil, but was used for a period of forty-seven years.

In 1912, Sam Spitz was a shipfitter at the naval shipyard, near Oakland, California, when he announced that he had invented an "aerial dial" that could see through dark or fog by "Stygian rays." The following year he told reporters that he had perfected his invention, which he named the spitzascope. He showed a reporter how the modern marvel could, from the inventor's basement workshop, project an image showing every movement of ships in the harbor. It could see anything within two miles, "through the darkest night and fog." He said that the US military had tested the device and offered to make him wealthy in exchange for the rights, but Spitz wanted his invention to benefit humanity, not be an engine of war, so he spurned the US military's offer, much as he had rejected an offer of a million dollars from the British government.[60]

Spitz organized the Wireless Spitzascope Company and sold stock. He leased the rooftop of a building in downtown Oakland, where he strung wires and antennae, and gave night demonstrations of the spitzascope. But some shareholders grew suspicious, and Oakland police detectives investigated. Spitz gave them his rooftop demonstration, showing how his invention, aimed this way, then that, could see through the darkness in all directions. The demonstration was convincing, until detectives noticed that where a new six-story building sat, the image on the screen showed only two stories, and still under construction. They arrested Spitz and found that his spitzascope contained a panoramic series of photographs that were several months old.[61]

Sam Spitz bluffed it through. He got out on bond and arranged for unhappy investors to drop their lawsuits. Spitz insisted that his invention was real, if not quite perfected.[62] The nationwide publicity given to his spitzascope swindle would have ended the career of a less tenacious salesman, but Spitz proved a master at promoting new inventions, and publicized himself as "the Edison of the West."

In 1919, he announced the invention of the marimeter, which used sound echoes to detect the depth to the sea floor. In later years he would claim that his invention was secretly used by the US Navy. In truth, the Navy inquired about the marimeter, but Spitz replied that it was still in the testing stage. The Navy developed echo-location without help from Samuel Spitz.[63]

Spitz did not serve in World War I, but by 1927 Spitz was telling reporters that he had been a lieutenant commander in the US Navy.[64] Like his military service, his education grew in retrospect. In 1914, he was a self-educated inventor, but by 1927 he listed previously unmentioned degrees: an electrical engineering degree from the Franklin Institute in Philadelphia, a bachelor of science degree from American Technical University, and postgraduate work at either the University of California or the University of Chicago.

In 1933, Spitz invented the "Spitz flight recorder," which displayed the location of every airplane in flight on a map of the United States. He said that it worked by detecting propeller noise up to thousands of miles away. This was an obvious fraud, but it made a good news story.[65]

Spitz announced his petrolometer oil finder in 1926. He said he had a perfect record of finding oil in nineteen attempts in the Los Angeles oil fields, and promoted his latest predicted oil field near Ogden, Utah. Spitz would drive a six-foot pipe into the ground, bury a microphone next to the pipe, then fire a pistol down the pipe, and listen for reflections picked up by the microphone. Spitz lied that the only thing that would reflect sound waves in the subsurface was an oil sand, so if the operator heard reflected sound waves, there must be oil beneath.[66]

This method had similarities to the seismic reflection method then coming into use, but the Spitz petrolometer could not work. Many geologic boundaries reflect sound waves, not just oil, and a pistol shot is far too weak an energy source for the purpose.[67]

In 1926, his petrolometer endorsed a drilling location at San Juan Capistrano, California, and two more at Encinitas. Shares in all were promoted through newspaper advertisements, and all were dry holes. Spitz promised oil at another San Diego County location in 1929, and two more in 1930—again all dry holes. In San Diego County alone, Spitz was zero for six.

Spitz marketed his petrolometer in Colorado. Although the account written by a professor at the Colorado School of Mines referred to the inventor as "Mr. Buzz," and his device as the "Buzzascope," the description leaves no doubt that it was Spitz and his petrolometer. Spitz interested some Denver oil men in the device, but when they had him test his device on a known oil field north of Denver, it failed. Spitz agreed to explain his invention to the professor of geophysics but then he didn't show for the appointment.[68]

By 1928, Spitz was active in Texas and Louisiana, claiming more than a hundred successful oil discoveries with only a few failures. Like many doodlebugs, Spitz's invention was applied to the wildest of wildcats; and like most doodlebugs, it failed. His machine predicted oil at both of Rev. David Olson's locations in Oregon in 1926, but the drill proved him wrong. In 1930, he picked a drilling location near Yakima, Washington, and although rumors circulated of a 1,000 barrel-per-day discovery, it was a dry hole. A petrolometer location in the Mojave Desert of San Bernardino County, California, was a dry hole. In 1940, his machine predicted oil at a location in central New Mexico, a location also endorsed by psychic Edgar Cayce. It proved another dry hole. Spitz claimed to have been hired by the famous wildcatter Tom Slick, but there is no record that he worked for Slick, and Spitz made the claim only after Slick died.[69]

The Southwest Research Institute of San Antonio, Texas, studied the petrolometer, and included its findings in an article in *Geophysics*. Neither Spitz nor the petrolometer is named, but the text describes what could only be Spitz's invention. The article described the electronic equipment as "crude and inappropriate," and the method as "pseudoscientific."[70]

The Spitz petrolometer seemed to have receded into well-deserved obscurity until 1955 when oilman George Ankarlo began using it to explore for oil in Illinois. He spent some months driving about Illinois and setting up the petrolometer to test the properties of hopeful farmers.[71]

Despite its inability to find oil, the Spitz petrolometer had remarkable longevity, and was used during a span of forty-seven years. In the end, Spitz's most successful invention was himself. Shipfitter Sam Spitz reinvented himself as Dr. and Lt. Com. Samuel Spitz, inventive genius. Samuel Spitz died in 1962 at age seventy-seven.

THE TRUMBULL AFFINITY INSTRUMENT

M. C. Trumbull wasn't an inventor or scientist. He was a manufacturer of barrel staves. But after an unsuccessful fling in oil wildcatting, he invented his Trumbull Affinity Instrument to avoid dry holes. Whatever it was, it became one of the most widely used oil doodlebugs.

Trumbull started as a lumber dealer in upstate New York. He followed the business to Arkansas and made barrel staves before moving his operation to Mansfield, Louisiana.

Oil drillers began hitting big around Mansfield, and in 1919 Trumbull joined other businessmen in Mansfield, Louisiana, and Pittsburgh, Pennsylvania, to form the Pittsburgh-Louisiana Oil Company. They sold shares through display ads in the Pittsburgh papers, but the drill rigs produced only dry holes for Pittsburgh-Louisiana, and by the end of 1920 the company had burned through its capital.[72]

Trumbull built a machine to avoid future dry holes. With help from a chemist cousin, Trumbull filled a container with his secret mixture of hydrocarbons that had an affinity for petroleum. Wires from the sample container connected to two iron stakes driven into the ground. When over an oil reservoir, the vibrations from electrons in the petroleum caused the electrons to vibrate in the instrument sample. The vibrations were amplified, and the output read on a meter on the face of the instrument.[73]

M. C. Trumbull first popped up with his affinity instrument in December 1921 in the oil boomtown of Mexia, Texas. Trumbull trained others in the mysteries of his invention, and by 1922 he had granted nine companies twenty-five-year licenses to use his invention. Trumbull travelled the country, checking the findings of his operatives.[74] Trumbull began extensive work in Kansas, and moved to Hoisington, Kansas, in 1923.

As long as he stayed close to known oil fields, Trumbull and his instrument did pretty well. But in the mid-1920s he began oil prospecting in places far from known production. In 1925, Trumbull invested in an oil test well in central New Mexico, but it was a dry hole. Trumbull picked a location five miles east of Lincoln, Nebraska—another dry hole.[75]

Trumbull's most high-profile failures were his dry holes in Arizona. In the mid-1920s, Trumbull and his affinity instrument predicted that oil would be found in at least eleven wells, two of them he drilled himself, but they were all dry. No oil has ever been found in Southern Arizona, but numerous oil tests were drilled, many—if not most—endorsed by one or another doodlebug operator. Besides Trumbull, doodlebuggers Boyd V. Lind and J. S. Gandy, both of California, and William Sharpe of Denver all looked for oil in Southern Arizona.[76]

M. C. Trumbull moved to Bradford, Pennsylvania, in the early 1930s, and directed his affinity instrument to finding oil and gas in Pennsylvania and New York. The last well drilled on the basis of the Trumbull affinity instrument appears to be a pair of successful gas wells in Wayne County, New York, in 1933.[77]

A. J. P. BERTSCHY

Adolph J. P. Bertschy was an inventive wizard, a self-taught tinkerer of the Edison type. His skill in making things work, from watches to automobiles to telephone systems, was always in great demand. His last great endeavor was to find oil with his oil-detecting machine, but it was perhaps his only invention that didn't work.

The son of French immigrants to Illinois, Bertschy dropped out of school at age ten. He burned through many careers: watchmaker, auto mechanic, electrician, telephone system manager, general manager of three automobile factories, metallurgist, and aircraft designer. It was not that he wasn't good at whatever he did, for everyone agreed that he was brilliant, and he mastered each new career with startling speed. Yet he never really prospered because he was always looking for the next technical challenge.

Bertschy rocketed through a career with the phone company in Illinois in the 1890s. Starting as a posthole digger, he quickly rose into management, and in 1904 invented a telephone relay amplifier. Bertschy quit the phone company to be an automobile dealer.[78]

Bertschy's auto dealership thrived because he was a whiz at auto maintenance. Fixing cars inspired him to make cars. The proposition wasn't as unlikely in 1907 as it seems today. The *Automobile Trade Directory* for 1909 listed 170 manufacturers of gasoline-powered "pleasure vehicles." From 1907 to 1911 Bertschy started and crashed three automobile manufacturers. His cars were praised as "masterpieces," but Bertschy's brilliance was too late—Ford rolled out his Model T in 1908, and from then on it was mass produce or die. No longer could a backyard tinkerer like Bertschy get by making a few custom cars.[79]

His first car factory was in Reno, Nevada, but Bertschy was distracted by inventing new processes to treat Nevada tungsten ore. The Nevada Motor Car Company lost money, and he sold the company to investors who moved the factory to Council Bluffs, Iowa. In 1911, Bertschy started another auto plant in Omaha, Nebraska, which also failed.[80] For all his technical talent, he failed because day-to-day business details could not hold his interest. It was the technical challenges that drove him.

In 1917, Bertschy patented a new process for hardening steel. In 1918 he was designing airplanes for the Army Air Corps.[81]

By the mid-1920s, Bertschy was still optimistic and enthusiastic, and running a machine shop in Omaha, Nebraska. Always looking for new worlds to conquer, he discovered oil exploration.

In 1926, Bertschy announced that after ten years of work, he had perfected an oil detector he called the Electrostatic Balance, or oil compass. No one knew how it was supposed to work, and he never applied for a patent. Bertschy claimed that his machine had resulted in the discovery of the Saginaw oil field in Michigan, and the oil fields near Russell, Kansas.[82]

He used his invention to locate a supposed oil field near Campbell, Nebraska. No one had yet found oil in Nebraska, but his enthusiasm and undoubted brilliance convinced the people of Campbell to raise $70,000 to drill Bertschy's location. Newspapers across the US printed a picture of Bertschy crouched next to his electrostatic balance, with a photo caption that sounded as if the oil were a sure thing and drilling the well just a formality—which was no doubt the way Bertschy thought of it.[83]

Bertschy's confidence seemed to be vindicated in August, when the well had an oil show at 2,605 feet. He announced that the well would make twenty-four to thirty-five barrels per day, and Nebraska newspapers hailed the state's first oil well. But when the company tried to produce oil, no oil flowed into the borehole. They deepened the hole but ran into drilling problems that ate up the remaining company funds. The drilling rig burned down in July 1927, when the well was down to 3,500 feet. The company vowed to rebuild and resume drilling, but never did.[84]

While the well at Campbell, Nebraska, was drilling, Bertschy went oil-hunting in South Dakota at a well at Standing Butte in central South Dakota. A. J. P. Bertschy studied the location with his electrostatic balance and pronounced it a dry hole. The place to drill, he said, was in the South Dakota badlands just east of Rapid City. Bertschy was right that the Standing Butte well would be dry, but his own well in the badlands was also dry.[85]

In 1932, Bertschy's machine convinced him that Omaha sat atop immense deposits of oil and gas. But, as before, he was unable to find with a drill the gas he found with his electrostatic balance. By the time Adolph J. P. Bertschy died in 1933, few took his oil-hunting gizmo seriously.[86]

LEE BOWMAN'S ATTRACTOMETER

You might think that an Attractometer is a dating app for your phone, and maybe it is. But years ago, the attractometer was an oil-finding machine created by Los Angeles inventor Lee Bowman. It was one of the longest-lived doodle-bugs, and Bowman used it for a period of nearly forty years, despite his bad habit of leaving it in his car to be stolen. The attractometer was one of his many inventions—some apparently real, and some highly improbable.

Levi Mack Bowman was born in Mississippi in 1885. His family moved to Texas, and Levi quit school by age fourteen to work as a janitor, but he had too agile a mind to stay a janitor. In 1913, at age twenty-seven, Bowman invented a pin holder, and filed for his first United States Patent. That patent, and six more, would be granted to him during the next ten years, for such disparate inventions as an electric pistol and a burial casket. Levi began calling himself

"Lee" Bowman, not knowing that a popular Hollywood actor would also use the same name.

One invention he never patented was his oil-finding machine, the attractometer. He never revealed the workings of the attractometer, which was a box mounted on a tripod. The outside of the box featured two thermometers, four gauges, seven dials, and an eight-day clock. Bowman said that he invented it in 1925, but it first came to public notice in September 1936 when Bowman reported to Los Angeles police that thieves had broken into his car and stolen his attractometer, "the only one of its kind in the United States," worth $2,000.[87]

A month before someone stole Bowman's attractometer, promoter Chester W. Colgrove began running newspaper ads offering use of the machine for finding oil. After it was stolen, the ads capitalized on the publicity, and reported that inventor Bowman was working feverishly to build another one, as the waiting list of clients grew.[88]

Colgrove moved his operation from Hollywood to Reno, Nevada, but continued to sell California oil properties. The US Securities and Exchange Commission obtained an injunction in 1938 against Colgrove for selling shares in a Sacramento Valley drilling project using doodlebug results. But this hardly slowed Colgrove, and four months later, New York blocked him from doing business in that state, after the state attorney general received a circular from Colgrove promising that a $250 investment in California oil land could return $60,000.[89]

Colgrove's apparently last and very ill-considered drilling project was the C. D. Murphy Jr. #1 State, a southwest New Mexico wildcat, located far from any good reason to drill. The well was plugged as a dry hole in 1951.

Bowman was arrested in Texas in 1937 on the charge that he had swindled a California rancher out of $4,000 in exchange for 10 percent of the profits from a survey of oil fields in Texas and Louisiana.[90]

In 1939, three men asked Bowman to invent a way to counterfeit money. Bowman alerted the FBI, who arrested the men. Three months later, fifty-three-year-old Bowman alerted police that his twenty-one-year-old wife had gone to a movie theater and never returned. He feared foul play from the counterfeiting ring. The search was called off when her parents reported that she was fine and was staying with friends.[91]

Lee Bowman patented a number of genuine inventions, but late in his career his inventions became less believable. During World War II, Bowman claimed to have invented an engine fueled by carbon dioxide. In 1954, Lee Bowman supposedly made a working perpetual motion machine, a series of wheels with permanent magnets spaced around the rims, which spun forever, as each magnet attracted and repulsed those on the adjacent wheel, and in turn was attracted and repulsed by the other magnets.[92] Legend has it that he let the motor run for a year, then dismantled the motor.

The attractometer made news again in 1951 when Lee Bowman reported that someone had broken into his car, this time in Shreveport, Louisiana, and stolen another attractometer, now worth $25,000.

The journal *New and Unusual Methods of Petroleum Exploration* asked Bowman for more information on the attractometer, and in 1951 received his brochure of unbelievable claims, which it quoted extensively. The journal closed: "it is unnecessary to inquire further."[93]

In the late 1950s Bowman formed the Geophysical Survey Syndicate to promote his attractometer and advertised in oil and gas trade journals. His ads proclaimed: "Science has overtaken the 'dowsers' and the 'doodlebuggers'" and promised that the attractometer could make a graphical log of oil saturation versus depth before drilling a well.

A 1959 article on "unorthodox petroleum exploration methods" has a photograph of the attractometer. The photo shows a box mounted on a tripod, with about a dozen dials on the sloping front face of the box. The author of the journal article classed the device among the "magical methods."[94]

By 1960, Bowman had two versions of the attractometer: his old-fashioned tripod-mounted gizmo, and one mounted on the dashboard of his car, so that he could watch the petroleum-indicator needle as he drove. He used his machines to place a drilling location near Ellendale, North Dakota, for the Herman Hanson Syndicate. The US Securities and Exchange Commission stopped the syndicate from selling shares to drill wells based on the attractometer, noting: "there is no scientific basis for locating oil by such method."[95]

The last time the attractometer came to public notice was in 1964, when Bowman—for the third time—reported the attractometer stolen out of his car. This time he hadn't been careless. Someone broke into his locked car parked in his locked garage. He told police that the box was booby-trapped to explode and possibly injure anyone who tried to open it without Bowman's special key. Bowman offered $2,000 for return of the attractometer.[96]

Lee Bowman died in 1972. He is remembered for his permanent magnet motor that supposedly ran without fuel. His attractometer, whatever it was, has been long forgotten, although it no doubt worked just as well as his perpetual motion machine.

M. S. BLACKBURN, ABNER DAVIS, AND THE RADIOSCOPE

San Antonio inventor McMaster Sylvester Blackburn first made the news in 1911 when he designed an airplane and formed the Chickasha Aeroplane Company to manufacture his flying machines in Chickasha, Oklahoma. However, the company never got off the ground. Blackburn was back in the news in 1921 with his Radioscope oil detector. Like many doodlebugs of the early 1920s, Blackburn's device was inspired by the radio craze. With so many doodlebugs jostling for attention, Blackburn was fortunate to have his invention properly tested by drilling wells. Unfortunately, his sponsor was a scoundrel named Abner Davis.

Abner Davis tried everything and failed at everything he tried. He was clever enough, however, to always fail with other people's money.

Davis first popped up in 1910 as the president of the Night and Day Bank in Oklahoma. His bank lent money to land speculator I. M. Putnam, accepting stock in Putnam's land development companies as collateral. Putnam promoted a land boom at a site he proposed for the state capitol building seven miles northeast of the official state capital of Oklahoma City. Putnam's land bubble burst, which sent the Night and Day Bank into bankruptcy. Davis was convicted in 1912 for mail fraud and was sentenced to five years in the federal penitentiary. His conviction was overturned on appeal, however, and the government dropped the prosecution in 1918.[97]

The following year, Abner Davis entered the oil business. With large display ads in eastern papers, he invited people to send money to Abner Davis Trustee Plan, to establish an integrated oil company that would drill wells, pump oil, refine it, and sell the product at its own gasoline stations. To avoid the inconvenience of government regulations, Davis made himself the trustee, and as trustee, he could do as he wished with the money—and what he did with the money was lose it.

Abner Davis organized the Radioscope Laboratories around M. S. Blackburn's invention and used the radioscope to find drilling locations for

Davis's Millionaires' Clubs. The Abner Davis Millionaires' Club sold shares through large newspaper advertisements in October 1921, looking for one hundred investors with $250 each. The club would drill near Mexia, Texas, where the infallible radioscope had defined an oil field look-alike to the Mexia oil field. When it hit, the club would split the profits fifty-fifty with Davis. The well proved to be a dry hole.

Abner Davis bought a helicopter design, patent applied for (and denied), from an inventor named Hackenberger. Abner Davis formed the Heliplane Company of America and sold shares aggressively until the US Postal Service hit him with a fraud order. After five years as a fugitive, a federal marshal arrested Abner Davis in June 1935. Within three weeks, Davis pled guilty to fraud, was sentenced, and started to serve his ninety-day jail sentence. Despite the token sentence, that ended the criminal career of Abner Davis.

M. S. Blackburn dropped from the headlines. He continued to promote his oil-exploration inventions but had a difficult time without Abner Davis's happy hyperbole. In 1929, Blackburn announced that he had invented a gravity meter that worked by an entirely new principle. He was awarded a patent for the method in 1931, but it was never widely adopted. In 1932, Blackburn introduced the Blackburn radiographic method, which used variations in reception of commercial radio broadcasts to predict oil reservoirs. No further details were divulged.[98]

Blackburn spent the late 1930s and early 1940s using his radiographic method in California and Alberta. When he died, the work with his radiographic instrument was continued by his coworker Worthy Gee into the 1950s, so that the radioscope/radiograph was in continuous use for more than thirty years.[99]

THE HERMAN HANSON SYNDICATE

Herman Hanson was a pioneer settler of Turtle Lake, North Dakota. He thought of himself as an amateur geologist, although some of his ideas were not endorsed by most geologists, such as when he claimed to have discovered the fossil of a winged creature a quarter of a mile long.[100]

In 1923, Hanson and some neighbors asked North Dakota State Geologist Arthur Leonard to tell them if there was oil beneath their farms. Leonard answered that conditions were not favorable, but the "possibility" of oil remained.

Hanson brought George W. Perry of Los Angeles to search for oil at Turtle Lake. Perry had invented a Mineral Indicator, a metal cylinder about eight inches long and an inch and a half in diameter, filled with secret chemicals, and suspended from Perry's finger by a silken cord. To detect oil, Perry placed a small piece of chamois skin soaked with crude oil atop his finger, wrapped twelve times with the silken cord. If copper was the object of the search, the chamois skin was replaced by a penny. If copper were directly below, the cylinder would be still; if not, the cylinder would begin to swing in the direction of the nearest large copper deposit. Perry demonstrated, and the cylinder swung in the direction of the copper mines at Butte, Montana, more than 500 miles away.

George Perry searched over the area using the oil-soaked chamois and found a large area where the cylinder would not swing. Perry told them, "I predict that development will quickly prove these lands the greatest oil field the world has ever known."[101] Hanson and fourteen neighbors pooled their mineral rights into the Herman Hanson Oil Syndicate and raised money by selling shares. They wanted their fellow North Dakotans to share in the bonanza, so they marketed only within the state. State Geologist Arthur Leonard tried to dampen their enthusiasm: "I am sorry that you believe the ridiculous stuff contained in that report and by the time you have spent many thousands of dollars drilling for oil and finding none, you will realize what a fake the man was. His findings are almost too absurd to deserve notice."[102] But the organizers believed Perry's doodlebug, so the Herman Hanson Oil Syndicate drilled at Turtle Lake in 1929. It was a dry hole.

The syndicate still believed Perry's mineral indicator and thought that another well at Turtle Lake would find the oil. In 1937, they prepared another stock offering, but this time, they had to deal with the US Securities and Exchange Commission (SEC). The SEC recognized that George Perry's cylinder was a doodlebug and forbade the Herman Hanson Oil Syndicate from selling shares.[103]

The Herman Hanson Syndicate revived after oil was discovered in North Dakota in the early 1950s. The syndicate brought in California doodlebugger Lee Bowman. Bowman's oil locations turned into dry holes. Another doodlebugger named Edwards was also unable to find oil.[104] The company finally gave up in 1959.

Spanning thirty-two years, from 1927 to 1959, doodlebugs misled the Herman Hanson Syndicate into drilling twelve dry holes. Company officers were undoubtedly honest. They always described their doodlebuggers as geophysicists, and no doubt thought they were basing their drilling locations on the latest science.

GEORGE W. PERRY: LOOKING FOR OIL IN ALL THE WRONG PLACES

George W. Perry was working as a painter in Seattle when he first brought his mineral indicator to public notice in 1914. He offered to use the mineral indicator to find a legendary buried treasure in Kansas. Then he appeared in Montana in 1921, doodlebugging for oil. There are conflicting descriptions of Perry's mineral indicator, so perhaps it changed over time. The version used in New Jersey was a compass and a dial indicating depth, mounted on a brass arm attached to a tripod. From the tripod, a cylinder containing a secret mixture of chemicals hung by a silk string. To test for a specific substance, a piece of that material was placed on the indicator, causing the compass to point toward any subsurface deposits of the substance.[105]

George Perry moved to Los Angeles and threw himself into the oil boom going on. He took out newspaper display ads touting Perry's mineral indicator and naming recent dry holes that he had supposedly predicted, emphasizing the money companies could have saved had they heeded his advice.[106]

George Perry also prospected in Texas and Nebraska, and he maintained that his mineral indicator was infallible. When he stated that he would find an oil field for a fee of $600, the *Independent Oil and Financial Reporter* of Fort Worth challenged him to bet $40,000: the *Reporter* would pay if Perry found an oil field; if the location produced a dry hole, Perry would pay. Perry did not answer the challenge.[107]

Doodlebuggers often do reasonably well in oil-rich areas, where with a bit of common sense and luck they may stack up a record of success. But when overconfidence leads them into blank spots on the oil map, they court their own downfall. The places that brought Perry's downfall were New Jersey and central North Dakota.

In 1925, Perry went to North Dakota for the Herman Hanson Oil Syndicate. The company drilled Perry's location to 1,840 feet before running out of money.[108] The syndicate later drilled two offsets to Perry's location, to 6,600 and 7,222 feet, and still found no oil.

In 1924, George Perry toured the East Coast with his mineral indicator. He

Fig. 18. Drilled locations, all dry holes, based on dowsing and doodlebugging in Iowa, Minnesota, and Wisconsin.

examined an area in central New Jersey, near Lakewood, that for no good reason had been punctured by a series of dry holes during the previous ten years. First the Perron company drilled two dry holes from 1914 to 1916. Then New York businessman William Driver and five Rhode Island textile manufacturers spent 1919 to 1922 and $400,000 drilling three dry holes at Jackson's Mill on nothing more than Driver's hunch. But when Perry stood over Driver's third failure, abandoned at 3,500 feet, the mineral indicator told him that he only needed to drill deeper.[109]

William Driver dropped out, but the Rhode Islanders agreed to fund further drilling. The borehole had passed into granitic basement rock at about 2,700 feet, and a geologist would say that there was no point in drilling deeper, but the mineral indicator said otherwise. They reentered the borehole and drilled deeper. By 1927 the borehole was deeper than 5,000 feet, with no oil. The Rhode Island investors finally gave up in 1928.[110]

The area around Jackson's Mill still fascinated investors. Another group restarted drilling in June 1928 but gave up in 1929. Yet another group took over and drilled still deeper. The borehole was finally abandoned for good in February 1930. It was a dry hole and a highly publicized fiasco for George Perry and his mineral indicator.[111] George W. Perry dropped from newspaper coverage after 1928.

WHERE THE DOODLEBUGS ROAM

As geology developed in the early twentieth century, the science defined areas unlikely to hold petroleum, such as the upper Midwestern states of Iowa, Wisconsin, and Minnesota. Most wells in these states were drilled for non-geologic reasons, such as dowsing and doodlebugs.

So why no oil in such a large area? Much of the area is underlain by sedimentary rocks, but the rocks were never buried deeply enough to have the heat and pressure needed to generate oil—with the exception of deep Precambrian sediments in the Keweenaw Rift.

Wisconsin

In 1929, a group led by Belle (sometimes called Alice) Jacobi, a spry seventy-year-old, leased oil rights on 12,000 acres near Rhinelander, Wisconsin, and announced plans to sell stock in the new Wisconsin Petroleum and Development Company. But to sell shares, they needed the permission of the Wisconsin Railroad Commission.[112]

At the commission meeting in May 1929, when geologists from the state geological survey and the University of Wisconsin testified that the whole idea was a waste of money, Mrs. Jacobi scorned the scientists, and told them that her divining rod had told her that the oil was there. There was not much that geologists and Jacobi agreed on. She lived in Zion, Illinois, a town owned and controlled by the Voliva religious cult, to which Jacobi belonged. The Voliva cult didn't believe in evolution, but did believe that the earth is flat, and also in complete chastity, even between married couples.[113]

The Railroad Commission decided that Mrs. Jacobi and company were sincere and granted permission to sell a limited number of shares, to drill a test well. The company raised the money and began drilling in September 1929.[114]

Meanwhile, some promoters started a company with a deceptively similar name: the Wisconsin Oil and Development Company, and began drilling in the next county, near Antigo. Once the well was drilling, the company split their lease into five-acre parcels and sold the small oil leases for ninety-nine dollars each, with the company retaining half the proceeds of any oil or gas found. The Railroad Commission refused them permission to sell shares. Neither well found oil.[115]

Not Finding Oil in Iowa

Very little oil has been found in Iowa. The state has produced only about 500 barrels of oil, a volume smaller than that of the gasoline stored beneath the typical corner gas station. But lack of oil has not stopped frauds and optimists from wasting a lot of money drilling for oil in Iowa.

C. H. Woodward was convinced that his farm near Manilla, Iowa, had oil, because the same kind of grass grew on his farm that he had seen in the Pennsylvania oil fields. He drilled shallow wells around his farm, and in 1904 announced that he had struck oil at 200 feet. Woodward showed a sample of his oil, and said that Standard Oil had tested it, found it genuine, and offered him $75,000 for his farm. The truth soon came out, however, that some boys had poured machine oil and kerosene down one of Woodward's borings.[116]

A doodlebugger detected oil south of Clarinda, Iowa, in 1922. Once the well started drilling, the people of Clarinda would not give up. By 1929, the well was down to 3,021 feet. Drilling stopped when the money ran out, but the doodlebug always seemed to indicate oil within the next hundred feet or so. Clarinda businessmen organized a new company, took over the well, and resumed drilling in 1930. The drill reached a depth of 4,671 feet before money ran out the following year. The chamber of commerce asked the town children to dip into piggy banks and spend their pennies in deepening the well. Local people dipped into their savings once more and paid to have the well drilled down to 4,980 feet in 1932. By 1934, the well was at 5,285 feet, and still no oil. The last few feet were drilled by farmer W. C. Davison, who got down to 5,308 feet before he quit.[117]

As oil excitement was building in southwest Iowa in the early 1920s, a landowner brought in doodlebugger John Getz from Illinois. Getz advertised himself as a geologist and used a contraption one writer said resembled a Ouija board, which Getz claimed was 90 percent accurate. He located an oil field extending more than 200 acres near the town of Humboldt, Iowa, with veins of oil at depths of 200 feet and 600 feet, with the best vein a thousand feet below the surface. Townspeople formed an oil company, and drilling began in October 1923.[118]

By May 1924, the well was down 800 feet, and the company was getting nervous about missing the two shallow veins predicted by Getz, so they called in another "oil expert," who could feel the "vibrations of oil." This expert told them that the well missed the shallow veins because of a fault but that the well should intersect the main vein at 1,000 feet. Reassured, the townspeople kept drilling, and didn't give up until they reached 1,005 feet without a trace of oil.[119]

Minnesota

When water dowsers dabble in oil, the results are usually bad, and the experiences in Minnesota proved no exception.

Jacob Berger was a Methodist minister and amateur oil dowser. He started oil dowsing in the late 1870s and had oil-dowsed in Texas and California before he retired in Faribault, Minnesota.[120]

By 1924, Berger was using a V-shaped fork with a series of wires connecting the two forks. Every couple of years from 1924 until his death in 1931, Berger announced another oil pool here and there in southern Minnesota, at Fairmont, Truman, Winnebago, and Lake Lillian. Local investors put together $20,000 to start drilling at Lake Lillian in August 1926. By January 1927 the drill was down to only 400 feet and was abandoned. Despite the dry hole and the negative opinion of the Minnesota Geological Survey, the following year a state senator proposed a bill directing the state survey to determine whether or not there was oil at Lake Lillian.[121]

The idea that there was oil near Lake Lillian led farmer Lester Case to hire local dowser Henning Nelson. Nelson walked the farm with his copper dowsing rod and found oil. Nelson's reputation for honesty and dowsing was enough to inspire sixty Lake Lillian businessmen to invest a thousand dollars each. Drilling began in January 1960 but was abandoned as a dry hole.[122]

Retired farmer and long-time water dowser Leonard Lenz took his copper L-rods to western oil fields and taught himself to dowse for oil. Upon his return, he found a number of supposed oil fields in southwest Minnesota and organized the Southwest Minnesota Oil Exploration Company. His dowsing reputation was so high that he found twenty-one people to back him with hundreds of thousands of dollars. The company leased about eighty square miles of land in eleven counties and in 1981 began drilling just north of the Iowa state line. The well hit Precambrian basement at 535 feet but drillers did not give up until they had reached 1,145 feet.

There is a remarkable thought-chasm between geologists and ordinary people concerning oil-dowsed locations. People were much more impressed by seeing copper wires dipping in the hands of an honest old farmer than they were by the unanimous negative opinion of oil-industry professionals and geologists of the Minnesota Geological Survey. Normally level-headed people lost large sums of money drilling where the rod dipped, despite scientific evidence that the area was barren of oil.

Possibility of Real Oil in the Upper Midwest
Despite all the false hopes and dry holes, a large oil or gas field might yet be discovered in the upper Midwest. A deep geological structure called the Keweenaw Rift, or Mid-Continent Rift, heads southwest from Lake Superior, passing under Wisconsin, Minnesota, and Iowa. The rift is deep and sparsely drilled, but someday it may yield oil or gas in Wisconsin, Minnesota, or Iowa.

THE SEC DECLARES WAR ON DOODLEBUGS
William Guest loved his nickname: Doodlebug. It summed up his profession and his claim to fame. But his name ran head-on against the law when the US Securities and Exchange Commission declared war on doodlebugs.

William M. Guest grew up in Detroit. He left school after the fifth grade for factory work, until his older brother moved to South Carolina, and William followed. He had learned to dowse at age eleven in Detroit and convinced the owner of mining property in South Carolina that he could dowse for tin. In 1917, at age thirty-one, Guest became a professional mineral dowser, and he began calling himself "Doodlebug" Guest.[123]

In 1921, Guest appeared in Los Angeles with a radio oil finder, and advertised himself as a consulting geologist. Guest hunted for oil in unlikely places. He predicted that wells drilled in South Pasadena would produce 3,000 barrels per day. The promoters printed Guest's baseless and wildly optimistic report in large newspaper ads, beneath the headline "Fortune Knocking at Your Door," to sell one-tenth-acre parcels. Another ultra-wildcat he endorsed was a well at Rancho Cucamonga, California, where the driller reported strong shows of oil and gas—which is doubtful. Famous oil journalist Ellwood Munger called the location "the last place a reputable geologist would put it." William Guest's most memorably wrong oil prediction was when he stood on top of Signal Hill, California, and declared that there was no oil there.[124]

William Guest moved his oil-seeking to Nevada. After two weeks of doodlebugging around Fish Valley in Nevada, he advised the Fish Lake Merger Oil Company to abandon the dry hole they were drilling, and drill at a spot selected by Guest's doodlebug. The Fish Lake company was eager to sell more shares, so they shamelessly hyped Guest's credentials. A company spokesman said of Guest: "[H]e has been successful in locating more gushers in Southern California than all other experts combined." Guest announced: "I am positive the Merger will have a gusher between 2,200 and 2,500 feet."[125] It was another dry hole.

William Guest announced in 1925 that he had a "Super Doodlebug." A news photograph shows a fancy metal Y-rod with a spring on each arm and on the main stem was a cylinder containing the secret substance, which he said included $5,000 worth of radium. Guest said that his doodlebug predicted oil with greater than 99 percent accuracy. He sometimes signed his letters "Super Doodle Bug Wm. Guest."[126]

Guest worked mostly in Texas, but he said that his doodlebug had detected giant oil deposits in unlikely places, including Connecticut, Delaware, Georgia, Idaho, Maine, Massachusetts, New Hampshire, North and South Carolina, Oregon, Rhode Island, Vermont, Wisconsin, and Washington state.[127]

> "I invented it. I worked at it for 35 years and just got it five years ago. Scientifically it's called a radiologometer. Familiarly, it's known as a doodlebug."
>
> —William M. Guest[128]

"[S]cores of men go about the country with queer oil finding devices making the foolish and impossible claim that they are able to locate oil in place. These devices are commonly called 'doodle-bugs.'"

—Lloyd N. Nash, SEC examiner[129]

Guest moved back to Detroit in the late 1930s and, together with stockbroker William B. Burbank, formed Producers Associates, Inc. They sold $43,000 in shares in oil-drilling projects in Michigan and Ontario, based on the locations chosen by Guest's super doodlebug. As usual, Guest's results didn't match his promises, and they were dry holes—every one of them.

The SEC harvested a great deal of positive publicity in 1935 when they halted share sales in the La Luz Mining Corp. The La Luz company had a gold mine which they claimed was extremely valuable based on uncorroborated indications from a doodlebug.[130] The SEC followed their success in the La Luz case by going after many oil and mining companies that used doodlebugs. Guest's very nickname, "Doodlebug," was a red flag to the SEC.

In September 1938, Guest faced off in court against SEC lawyers. The judge was unimpressed by Guest's doodlebug demonstration and issued an injunction preventing Producers Associates from selling oil and gas interests. The SEC then took a criminal case against Guest and Burbank in 1939. Guest couldn't post the $5,000 bond, so he waited in jail. Burbank remained in Canada until he surrendered to American authorities three months later. Burbank and Guest saw that the case was going against them, so they made deals with the prosecutor. They both pled guilty, and the judge sentenced Burbank to eighteen months and Guest to a year and a day. This ended the career of William "Doodlebug" Guest.[131]

SOME OTHER "BUG MEN"—AND A WOMAN
Benjamin F. Fulton

Ben Fulton was a grocer in Portland, Indiana, until the gas-drilling boom came to Indiana in 1889, and he became a drilling contractor.[132]

Fulton announced in 1894 that he had invented an electromagnet that would find underground gas. His gas finder did not attract much attention until he discovered a new gas field west of Portland, Indiana, in 1897. By 1897, his electromagnet had been replaced by a pair of aluminum tubes the size of fountain pens, wrapped in rubber. Fulton would attach the tubes

Fig. 19. L. V. J. Kimball came to oil doodlebugging after years of experience doodle-bugging for minerals. (*Pharos-Tribune* (Logansport, IN), August 24, 1896.)

to the ends of the prongs of two flexible metal rods joined at one end to form a "v" shape. He would grip one tube in his teeth and hold the metal rod so that the second tube pointed forward. The tubes were filled with chemicals having an affinity for the searched-for substance, whether oil, gas, gold, or silver.[133]

Fulton's wells often outperformed others in the area. By July 1897, he claimed to have made more than five hundred well locations in northeast Indiana and adjacent Ohio. His fame spread after he drilled a number of high-volume gas wells, and he was invited to West Virginia, where he drilled a field-discovery well in Upshur County. By 1900, Ben Fulton was president of the company that had hired him as a drilling contractor, the Citizens Gas and Oil Mining Company, of Portland, Indiana. He was still active in oil drilling in 1908. He died in 1911.[134]

Professor L. V. J. Kimball, the Magnetic Man

In 1901, Professor LaVergne J. Kimball showed up in Long Beach, California, promising to find oil with his own instrument. Although new to oil, Kimball brought along a decade of experience doodlebugging for minerals in Nevada, California, and Colorado.

The "professor" had no university background. Kimball had been a painter by profession who grew up in Michigan before moving to California. In 1891, at age forty-fix, Kimball popped up in Colorado during the Cripple Creek mining boom as "Professor" Kimball with his Electro-Magnetic Mineral Indicator.

As did many early doodlebugs, Kimball's mineral indicator looked a lot like a traditional forked dowsing rod. It was made up of a pair of heavy copper wires connected at one end to a bullet-shaped element he called a magnetic-electric battery, the construction of which was secret. He had a different battery for each substance, and he said that it would detect oil not only in the ground beneath the instrument, but also any oil within twenty-five miles. Although called "magnetic-electric," Kimball also said that his batteries were "purely magnetic and not electric." He said that it worked for him because he was also magnetic, explaining: "some people are electric and others are magnetic."[135]

As an oil explorer, Kimball continued to use his mineral indicator, but he kept it mostly out of sight. He changed his title from "professor" to "scientific geologist," and said he was a graduate of the University of Texas. He picked well locations in a wide swath of the United States: California, New Mexico, Colorado, Kansas, and Texas. He seems to have failed wherever he went, even though he said that he had chosen more than three hundred oil well locations, without a single failure.[136] LaVergne J. Kimball kept doodlebugging until his death in 1921.

An Oil Well or Your Money Back

The Oil Finding and Developing Company made an unbeatable offer: strike oil, or your money back. Not just a refund of the fee, but also drilling costs. They could make this offer because they had exclusive use of the Canfield oil finder, and the services of inventor George B. Canfield (no known relation to oil dowser Isaac Canfield).

Canfield was an Illinois farmer who invented an oil-finding machine. The invention was never described in press accounts. He explained only that oil veins are charged with energy, "probably magnetic," which made his machine work. Canfield said that he had tried his oil finder successfully at oil fields in Illinois, Indiana, Pennsylvania, and Texas.

Some Chicago businessmen created the Oil Finding and Developing Company and offered a unique deal: the company received $1,000 per location, plus 25 percent of all the oil produced. If the well did not find oil, the company would not only refund its $1,000 fee, but also repay the cost of drilling. A company officer explained: "Our instrument is reliable and as infallible as any scientific instrument in its operations. It never makes any mistakes."[137]

George Canfield had used his oil finder to trace thirty-five underground streams of oil flowing out of Wisconsin, crossing Illinois and into Indiana. Each stream was a swarm of veins, he said.

The Oil Finding and Developing Company contracted with oil companies drilling at Mattoon, Illinois, and Muncie and Parker City, Indiana. The Oil Finding and Developing Company earned newspaper headlines during January through March 1904, then disappeared, gone from newspaper headlines, as if it had never existed. We can only guess that George Canfield's oil finder ran into bad luck, and the Oil Finding and Developing Company went out of business. Canfield persevered, however, and in late 1904 was drilling near his farm in Ogle County, Illinois. He found no oil and died in 1907.

Finkenbiner's Doodlebug

John S. Finkenbiner of St. Louis quit a career as an executive with the Singer Sewing Machine Company to try his luck in oil, but he had nothing but bad luck drilling in northeast Indiana. In 1903, after two years and three dry holes— but with strong shows of oil—he brought in a man from Chicago with an oil-finding instrument. Two more dry holes were the result.[138]

Finkenbiner returned the following year with an oil finder of his own invention. He used it to find drilling locations in eastern Indiana and southern Arkansas, but apparently with no more success. He dropped out of oil finding and turned his oil locator into a mineral ore finder. He worked in the mining industry until his death in 1909.[139]

C. L. Cofer and his Terrestrial Wave Detector
C. L. Cofer was a rancher near Red Bluff, California, with a reputation as a water dowser. In 1909 he announced that he had invented an electric Terrestrial Wave Detector, which could detect underground water, oil, and natural gas. It was a supposedly scientific device, but how or why it worked, he never seemed to explain, and no one seemed to care. All people cared about was that he could find things underground.

Cofer benefitted from initial successes in 1909, but later dry holes drained the enthusiasm. He said in 1911 that he had located a large gas deposit next to the city of Oakland and a huge oil field near Richmond, California, but drilling found nothing. In 1912, he found a supposed oil field in northern Napa County that resulted in another dry hole. Nevertheless, Cofer continued to find work locating oil and gas wells.

Cofer was last heard from in 1924, when two wells for which he predicted oil were abandoned as dry holes. Fellow doodlebugger David Olson brought Cofer to Oregon to pump some enthusiasm into investors for wasting more money on Olson's dry holes. Cofer obliged by saying that his terrestrial wave detector had determined that the location at Eugene was underlain by three "Gusher Oil sands" that would produce millions of barrels of oil. The well was abandoned without finding oil. Also in 1924, a well in California that Cofer said would hit oil at 910 feet was abandoned at 1,035 feet. These were Cofer's last known failures.[140]

Lucretia Campbell's Synchronal Compass
Miss Lucretia Campbell lived on a farm near Chanute, Kansas, and in 1914 came forward with her invention, the Synchronal Compass, which detected oil deposits. The construction or operation of the compass were never described in newspaper accounts, but she used it to become a successful oil driller.

It might be assumed that Miss Campbell was a spinster who never left the family farm. In fact, at age nineteen she married John F. Cox, a clothing salesman, but after twenty-two years of marriage and four children, she divorced him and reclaimed her maiden name. The separation was not amicable. A newspaper reported that "[t]he divorce petition was one of the most sensational ever filed in Montgomery County."[141]

After her divorce, she moved back to the family farm that her father had homesteaded and told people that she was a widow. Campbell became a leading suffragette in Kansas, and lobbied state legislators to secure votes for women. Campbell also started drilling for oil. In 1911, the Topeka *Daily State Journal* called her "the oil magnate of the Humboldt district" and noted that state legislators accorded her and her arguments more respect because of her business success.[142] Kansas men extended the vote to women in a statewide referendum the next year.

Miss Lucretia Campbell wrote to the *Oil & Gas Journal* in 1914 that during the past few years she had tested her synchronal compass on oil fields as far south as Mexico and found it to be infallible in locating oil. She scorned men who said that only the drill bit could reliably find oil and challenged the *Journal*'s readers to put her to the test: "[I]t is a woman who now invades the sacred knowledge chamber of man and throws down the gauntlet. I stand ready to prove my statement."[143]

Her synchronal compass must have served her well because she kept drilling wells. The novelty of her gender drew publicity, and newspaper headlines proclaimed her "queen of wildcatters." When the *Chanute Tribune* referred to her as an "oil witch," then a common term for oil dowsers of either sex, she reminded the editor that no gentleman would describe a lady as a witch, unless the word is preceded by "be" and ends in "ing."[144] She apparently retired from oil drilling in 1922, and ran her farm until she died in 1954, at age eighty-nine.

Ramon Carrillo

Ramon Carrillo was from an old California family; his grandfather and namesake Don Ramon Carrillo owned huge ranches and fought against the Americans in the Mexican–American War. Grandson Ramon Carrillo worked as a car mechanic in Santa Ana, California, and tinkered with oil-finding instruments. In 1921, after nine years of experimentation, he took his doodlebug public and started predicting oil-well results.

In 1923 Carrillo demonstrated his oil-finding instrument to George Bentley, a businessman visiting from Hutchinson, Kansas. Bentley returned to Kansas and convinced two other Hutchinson businessmen to send for the doodlebugger. Bentley didn't mention that Carrillo had agreed to split his doodlebugging fees fifty-fifty. Carrillo arrived in Hutchinson and selected a spot north of town.[145] The well hit oil at 3,370 feet and became the discovery well of the Welch-Bornholdt field. The townspeople held a parade in Carrillo's honor. One of the investors owned an auto dealership and gave Carrillo his pick of a free Lincoln.

Carrillo and his wife moved to Kansas, but after doodlebugging a dry hole in Anderson County, Kansas, he returned to California in 1925. In 1929 he organized the Ramona Oil Company.[146] Carrillo's oil-industry fortunes went downhill, and by 1940 he was working in Los Angeles as an oil-field laborer. He had gone full circle, from mechanic, to petroleum engineer, to roughneck.

G. C. Wucherer and his Oil-Finding Instrument

G. C. Wucherer first used his instrument to locate drilling sites in Southern California, and he seems to have done fairly well—as long as he stuck to oil-bearing regions.

A realtor eager for an oil boom invited Wucherer to Arizona, where he outlined a huge oil field between Phoenix and Wickenburg, another oil field closer to Phoenix, and a large field near Casa Grande, Arizona. Wucherer's well northwest of Phoenix began drilling in early 1921, and in July he announced that the borehole had cut an oil sand, and that it was "a certainty" that the well would be an oil producer. But the well never produced oil, and drilling stopped in June 1923 at a depth of 3,350 feet. Like every other well drilled in Southern Arizona, it was a dry hole.

Wiley Post: Aviator and Doodlebugger

The pioneer aviator Wiley Post is today mostly remembered for crashing his plane on takeoff at Point Barrow, Alaska, killing himself and his sole passenger, friend and fellow Oklahoman, humorist Will Rogers.

Not too many people knew of Post's attempt to build a machine to detect underground oil. He conceived of a way to use magnetism and tested it by driving around over a known oil field. He at first thought that he had succeeded but found that his instrument indicated oil everywhere. He believed that it was too sensitive, and that it was registering the oil in his crankcase. Post thought that he was close to a solution, and worked devotedly on the invention, but finally gave up.[147]

DOODLEBUGS AROUND THE WORLD

DOODLEBUGS NORTH OF THE BORDER

The author of a history of the oil industry in western Canada described the differences between the oil industries of Canada and the US, noting—with perhaps a trace of self-congratulation—that "'doodle bugs' are practically unknown in Canada."[1] In fact, the Canadian and US oil industries were similar in their use of dowsing and doodlebugs. As in the US, Canadian doodlebugs were common.

Lyle Telford, Mayor of Vancouver

Dr. J. Lyle Telford was known in Vancouver, British Columbia, for his radio show offering medical advice.

In 1914, he invented an oil finder that resembled a small flashlight hanging by a cord. When held over oil, part of the machine would spin. A Vancouver syndicate sent him to Calgary to find oil-bearing land. To establish credibility, he handed a sealed envelope to the editor of the *Albertan*, containing Telford's prediction of a well being drilled. A week later, when the drillers of the Monarch well announced that they had discovered oil, the editor opened the envelope: Telford had called it correctly.[2]

The Canadian Pacific Oil Company took Telford to California's Central Valley, where he tested his instrument over about thirty drilled wells, and

Fig. 20. J. Lyle Telford had an extraordinary career as a physician, radio broadcaster, oil doodlebugger, and politician.
(Photo courtesy of City of Vancouver Archives.)

correctly read the results of each. He then predicted results for three planned wells and returned to Vancouver. All three wells came in as predicted. When Telford returned to California in 1915, he charged $500 for the first location, and $250 for each additional location.[3]

After 1915, Lyle Telford disappeared from the columns of oil news. Years later, his brother, the Vancouver dentist George H. Telford, entered the oil-locating business using an instrument similar to the one used by Lyle. In 1923, George Telford was busy picking drilling locations at the Sunburst field in Montana and was planning to travel to Mexico to evaluate 350,000 acres for some English capitalists. George Telford's oil-locating career was briefer than his brother's, for no mention of his oil-locating appears after 1923.[4]

Lyle Telford returned to his radio show, and in the 1930s became influential in the Co-operative Commonwealth Federation (CCF) political party. He was elected mayor of Vancouver, British Columbia, in 1939.

M. V. Bennett

M. V. Bennett was a Canadian immigration agent at Omaha, Nebraska, who perfected an oil-finding device in his spare time. In 1921, he convinced some Winnipeg investors that his oil finder had detected oil at Stony Mountain, north of Winnipeg.[5]

Geologists knew that Stony Mountain was a very poor place to drill for oil because of basement rock at a shallow depth. But the investors were convinced that Bennett's instrument was 100 percent accurate and formed the Stony Mountain Oil and Gas Company to prove it. Bennett became company vice president. The company started drilling the well in 1922, and as expected, drilled into granite at 1,000 feet, and was abandoned as a dry hole.[6]

The Watson Ethereal Wave Petroleum Magnet

Hugh Watson boasted thirty years of experience as a geologist in Canada, the United States, and Australia. He announced in 1922 that after fourteen years he had perfected the Watson Ethereal Wave Petroleum Magnet. The machine not only detected oil beneath it but also could detect oil four to ten miles away.[7]

Triumph Oil and Gas sold shares through newspaper ads touting Watson's invention as offering "the absolute certainty of getting petroleum." Triumph raised money to drill three "large lakes of petroleum" discovered by Watson along the Fraser River in British Columbia.[8]

In 1923, Triumph Oil sent Watson to find oil in Montana. Great Plains Oil hired Hugh Watson in 1925 to find oil on their 1,760-acre leasehold near Limerick, Saskatchewan. Watson found large oil-bearing structures, and Great Plains Oil used his report to sell shares in the wells.[9] Hugh Watson and his Ethereal Wave Petroleum Magnet disappeared from the news after 1926.

The French Marquis and Tele-Detection

In August 1940, an ocean liner from Britain docked at an unnamed Canadian port, with a cargo of refugees. Newspapers skipped the details so as not to tip off German U-boats. Among the refugees was Marquis Georges Paul d'Aigneaux, a resident of Paris before the Germans marched in two months before.[10] D'Aigneaux knew Canada, having lived there, and had made a career dealing in Canadian furs. But this time he came off the boat with an instrument that had nothing to do with furs: a doodlebug to find oil by radioactivity.

How d'Aigneaux got the doodlebug is unknown. He said that he knew and worked with the ultimate scientific power couple, Pierre Curie and Maria Skłodowska-Curie. D'Aigneaux went to Calgary, Alberta, and started marketing his Radio-Electro-Magnetic Teledetection for oil and minerals. A newspaper advertisement in the Toronto *National Post* from September 1941 said that his system could detect oil and gas more than a hundred miles away.[11] This appears similar to the doodlebug developed in France in the 1920s by Henri Moineau and a Mr. Regis.

D'Aigneaux was widely regarded as a faker, but in 1941, oil man Tom Keyes hired him to survey for oil and gas reservoirs in Saskatchewan, and d'Aigneaux outlined an area that later became a large gas field.[12]

The Marquis d'Aigneaux's mental state deteriorated, and he spent nine months in a Swiss mental hospital in 1947–1948. Upon his release, he returned to Calgary, where he died of heart failure on Christmas Eve 1948 at age seventy-four.[13]

Clarence Stork

Clarence Stork was a Saskatchewan farmer who helped organize the Co-operative Commonwealth Federation (CCF) and represented the CCF in the provincial legislative assembly from 1934 to 1938. He moved to Calgary and won a reputation as an effective oil finder in Alberta in the late 1940s, using a "doodlebug apparatus" whose construction he jealously guarded; he didn't want people to even see it. He went to Ontario in 1950, and in 1951–1952 drilled an unimpressive five oil producers and eleven dry holes. Clarence Stork was active in the oil business until his death in 1970.[14]

West Drumheller Oil Field

After the discovery of the Drumheller field in 1950, thirteen investors in the town of Drumheller, Alberta, decided to get in on the oil boom. They optioned some tracts six miles west of Drumheller field. Great Plains Development Company crossed the tracts with seismic surveys and found nothing, but the syndicate chose to instead trust Glen Phillips, whose electronic device had detected oil on the tracts.

The Mazel #1 well drilled down at Phillips' location in 1952. The well blew oil hundreds of feet in the air and was completed making 1,220 barrels of oil per day. Phillips doodlebugged more locations until the West Drumheller field had thirteen producing wells.

Innes Paterson's Three Doodlebugs

William Innes Paterson was a Vancouver oil doodlebugger who divided his time between western Canada and the United States. Despite the services of three doodlebugs of increasing sophistication, all his wells seem to have been dry holes.

Paterson was in the lumber business in British Columbia before joining the oil search. In 1912 he raised some foolish investment money and drilled at Pitt Meadows on the east side of Vancouver, at a spot selected by a California doodlebugger named Mr. Frank. The *Mining and Engineering Record* in Victoria, British Columbia, noted that "[t]he geological conditions at Pitt Meadows are absolutely opposed to the presence of economic oil deposits."[15] Frank had a history of drilling dry holes, and his location at Pitt Meadows was no different. The well hit an uncontrollable flow of water at 1,228 feet and was abandoned as a dry hole in 1913.

Paterson returned in 1914 and drilled another well 250 feet from the first, this time based on an instrument called a Mineometer, invented by chemist Samuel Scott of Tacoma, Washington. The effectiveness of the instrument depended on the secret chemical mixture with an affinity for crude oil that was placed at the tip of the dowsing rod. Not all found the mineometer convincing. The *Mining and Engineering Record* pointed out that the mineometer had already failed when it was used to search for silver and lead ore and failed again when it was used to search for coal.[16]

The company exhausted its funds, and Paterson sold more shares with newspaper advertisements encouraging the public to buy shares immediately because the price might increase at any hour. Vancouver newspapers cooperated by printing optimistic headlines such as "May Be Big Gusher of Thousand Barrels."[17] The second well was no more successful than the first and was plugged. Paterson had exhausted his credibility in British Columbia, so he went to Wyoming and Montana in 1916.

Paterson arrived in Nebraska in 1917, and by the following year he used an electrical instrument to trace a great underground river of oil running into Nebraska from the oil fields of Wyoming. He incorporated and sold shares in the Buffalo Oil and Gas Company to drill at Red Cloud, Nebraska.[18] It was another dry hole.

Paterson's electrical instrument—he called it an Oil Thermometer—was a metal cylinder hanging by a wire, which swung and spun in a particular manner when over oil.[19] Undeterred by his failure at Red Cloud, the following year Paterson sold shares in the Nebraska Land and Development Company to drill an oil field found by his oil thermometer near Beatrice, Nebraska. It was another dry hole. Paterson had discredited himself in Nebraska, but by then he had moved on to Kansas.

In 1920, Paterson followed his great river of oil into Kansas, where he dowsed a number of undiscovered oil fields. By this time Paterson had added an oil compass to his doodlebug arsenal. The oil compass was built for him by an unnamed Canadian university professor. Paterson described it as having a compass-like needle that wavered rapidly when over oil. The description sounds like a Mansfield Automatic Oil Finder, or similar device. He used the oil compass to find the oil, and the oil thermometer to define the exact edges of the pool.

Paterson formed the Indian Mound Oil Company and adopted the discreditable practice of drilling without sufficient funds. Paterson knew that once a well was started, he could sell more shares by spreading rumors of oil discoveries. He began drilling at Winchester, Kansas, in June 1921, but soon ran out of money—so he stopped drilling to raise more funds. Drilling restarted in November 1921, and a few days later hit a flow of gas. Paterson guaranteed that oil would be found just below the gas. A thin scum of oil in the borehole was exaggerated into an entire borehole full of oil—and Paterson sold more shares.

The editor of the Atchison, Kansas, *Globe* cautioned that finding oil with a doodlebug was not at all the same as finding oil with a drill rig. Some called the *Globe* a "menace" to prosperity for its warning. But the *Globe* was vindicated when the Winchester well drilled through the expected oil sand and found it full of salt water. Paterson was undeterred and promised that oil would be found in a deeper sand. The company ran out of money, and drilling stopped in December 1921. For some reason the Winchester well fascinated investors, who periodically reentered and deepened it. It was abandoned in 1927, then again in 1929 at 3,440 feet with the bottom in granite. In January 1931 the well was being deepened further.[20]

While the well at Winchester was being drilled off and on, Paterson was drilling a test well near Beattie, Kansas. By December 1921, there were two contrasting stories of the Beattie well. The editor of the Frankfort *Index* wrote that the well had found oil, but the company was keeping it quiet. The Atchison *Globe* was more cynical and dismissed the rumors. Other wells were drilled by Innes Paterson in Nemaha County, Kansas, and near Oak Mills, Kansas. They were all dry holes.

Between dry holes in Kansas, Innes Paterson found time to promote and drill a dry hole near Stonewall, Manitoba. He leased 10,000 acres in 1921 and started drilling in June 1922. He relied on his doodlebugs, and they were complete failures. Innes Paterson stayed in Kansas until 1940, when he moved to Alberta to promote oil prospects before he retired to Vancouver.

MAX STEINBUCHEL DOWN UNDER

Max Steinbuchel was a highly successful oil operator in Wichita, Kansas, but his Radiograph, so effective in finding oil in Kansas and Oklahoma, could not locate the more elusive Australian oil.

Steinbuchel was a big man, tall and heavy-set, with a craving for adventure. Born in 1902, he apparently lied about his age to enlist in the US Army during World War I. He was injured and in 1920, more than a year after the war ended, he was still recuperating in an Army hospital near Denver, Colorado. He spent some years as a sailor before returning to Wichita.

Max Steinbuchel became a construction contractor, then in 1929 he joined with an experienced oil man to drill a wildcat oil well in Rice County, Kansas. They drilled the discovery for the Schurr field then sold it to wildcatter Tom Slick for a quarter-million dollars. After that, Steinbuchel scored one success after another.

Steinbuchel turned up in Melbourne, Australia, in July 1938. He was on holiday, but he spent his time walking with a doodlebug over the Gippsland region of Victoria state, where the government was drilling a test hole. Steinbuchel was looking for oil—and he found it, or thought he did. Local newspapers publicized his findings.[21]

Australians were of course familiar with water and mineral dowsers, but only small amounts of oil had been discovered in Australia, and the country had no verified oil dowsers. Australian dowsers had tried to find oil, but only compiled a list of failures. One of the most highly publicized failures was at American Beach, South Australia, where in 1921, a company drilled hundreds of feet into schist on the word of a dowser who insisted that he detected oil at the location.[22]

As a successful oil driller, Steinbuchel commanded respect in an oil-thirsty nation. Steinbuchel said that he bought his radiograph doodlebug from a German scientist who had since died. He insisted that it was a geophysical device, and that he was a geophysicist, not a dowser, but government geologists recognized it as a dowsing-type instrument.[23]

By the end of 1938, Great Eastern Oil was moving its drilling equipment to one of Steinbuchel's locations. The Melbourne businessmen who ran Producing Oil Fields Ltd. were organizing an associated company, Producer's Oilwell Supplies Ltd., to promote Steinbuchel's oil locating. Producer's Oilwell Supplies Ltd. drilled four of his doodlebug locations—all dry holes—in the Gippsland Basin. His backers had not lost faith, however. He chose five locations in South Australia and began drilling near Mount Gambier in 1940. Steinbuchel also planned to drill locations in New South Wales and Tasmania.[24]

In March and April 1940, Max Steinbuchel became the center of debate in the Australian parliament. The Minister for Mines warned the public against investing in Steinbuchel-associated companies. Detractors said that he was a swindling charlatan and wanted him expelled. His six-month tourist visa had been extended, but the Australian government announced that it would grant no more extensions, and Steinbuchel had to leave by June 4, 1940. His supporters said that he was an honest and successful oil man and accused his opponents of being tools of powerful oil monopolists. Senator Arthur promised that if Steinbuchel were allowed to stay, and failed to find oil within twelve months, then Arthur would resign from the senate.[25] Senator Arthur wrote to the editor of the *Oil & Gas Journal*, of Tulsa, Oklahoma, inquiring about Steinbuchel's qualifications. The *Journal* confirmed by letter that Steinbuchel was a very successful oil operator. It recounted Steinbuchel's early field discoveries: Shurr, Raymond, and Isern, and his later operations in Oklahoma and Texas.[26]

In December 1940, Producer's Oilwell fell apart, as rival factions took turns locking each other out of the headquarters.[27] Steinbuchel left quietly and arrived in Hawaii in May 1940. Once back in Wichita, he immediately started leasing and drilling, as if he had just been on a brief holiday rather than gone for two years. Max Steinbuchel died in 1954.

EUROPEAN DOODLEBUGS

As in North America, the number of doodlebugs in Europe rose quickly in the early 1900s and peaked in the 1920s. US and Canadian doodlebug inventors were spread across a wide range of professions and education levels, but almost all European doodlebug inventors were from the educated class: priests and physicians in France, engineers in Germany. Perhaps because European doodlebug inventors were drawn from a smaller group, there seem to have been many fewer doodlebugs in Europe than in North America.[28]

French Doodlebugs

Pierre Estinès was a priest, doctor of mathematics and physics, retired college professor, and water dowser. In 1922, at age seventy-two, he announced that after twelve years of work he had perfected a machine that could infallibly locate oil fields and specify the depth. Estinès said that he had already found enough oil in the French Pyrenees to supply France's oil needs.[29]

Estinès would not patent his oil-finding apparatus, commenting that a patent would only make it easier for others to steal his idea. He would only explain that his machine worked by radiation. It was said to weigh 300 pounds and worked best between 2 a.m. and 5 a.m. In the span of seven years, investors drilled numerous wells on locations indicated by Estinès, on both the French and Spanish sides of the Pyrenees, but despite some premature reports of success, his invention did not find any oil.[30]

Henri Moineau was a physician and water dowser. In 1922, he and a Mr. Regis announced an invention to detect oil deposits, and they didn't even have to move their machine to the oil field. Their machine stayed in place and used radio waves to detect oil fields miles away—hundreds and even thousands of miles away.

From their laboratory in Puy-de-Dôme in central France, Regis and Moineau announced that they had discovered new oil fields in France, Italy, Germany, and Czechoslovakia. They even said that they had discovered a new oil field in the Rocky Mountains of the United States, a third of the way around the globe. The inventors said that the exact drilling locations would be determined by what they called aerial x-ray photography. During the next several years, Moineau drilled a number of dry holes. Geologist and oil historian Pierre-Louis Maubeuge gave Moineau credit for being one of the few doodlebuggers honest enough to risk—and lose—his own money.[31]

Pierre Guerenneur was a naval engineer who in 1928 invented a device to detect petroleum with radio waves. He detected sixteen oil fields in the Aquitaine Basin of southwest France, and received financial backing to drill four of them, but the drilling depth was limited to 350 meters or less. He found no oil, and drilling stopped by 1930. According to journalist Pierre Fontaine, the Standard Oil of New Jersey 1954 discovery of the Parentis oil field, the most productive oil field in France, was made by drilling deeper on one of Guerenneur's prospects from the 1920s.[32]

Guerenneur also used his oil finder in Morocco in 1933 and Algeria in 1946 but did not succeed in having his prospects drilled.[33]

Jean Jeannet, a worker in a Paris rubber factory, announced in 1930 that he had invented an infallible oil finder he called the radioscope. It was a glass tube filled with a mixture of oil and secret chemicals. The tube hung by a wire and would vibrate and bob up and down when near oil. Directly over oil, the tube would swing in a circle. It is not known if he got the chance to prove his invention.[34]

The Schermuly Polarizer: World Doodlebug

In 1916, German drilling engineer Philipp Schermuly applied for a patent on his latest invention. He already had a number of patents on improvements in drilling machinery, but this invention would carry his name around the world. The Schermuly Polarizer would become one of the most popular doodlebugs in the twentieth century and was used around the world for oil and mineral exploration. Today it is forgotten.

Calling it a "polarizer" suggests a more complex device. It was a mechanical spinner: a metal rod held vertically between the index fingers, with an arm extended out with a capsule at the end. Schermuly had about a hundred capsules for different substances or geological conditions. The rate of spin determined the distance and concentration. Schermuly patented the mechanical apparatus, but the contents of his many target-specific capsules he kept secret.

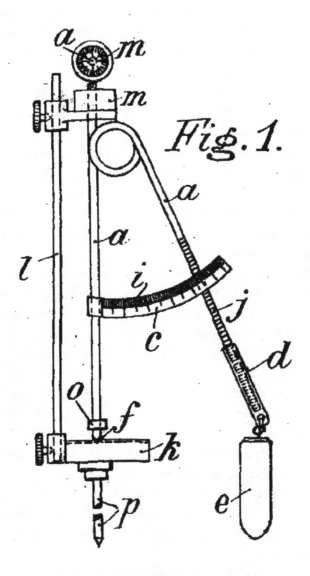

Fig. 21. Master engineer Philipp Schermuly invented the Schermuly Polarizer.

The original polarizer had the capsule at the end of a spring angled from the rotating axis. Later models replaced the flexible spring with a stiff arm and added features to measure the spin rate and the angle between the arm and the axis.

Schermuly was an acknowledged drilling expert and carried the respected title *oberingeneur* (master engineer). He was also a long-time dowser and acknowledged that his polarizer was based on dowsing principles. Schermuly said that dowsing was accurate only in locating water, but that his polarizer was 95 percent accurate in finding minerals and oil. In addition, he said that anyone could use the polarizer, not just talented or experienced dowsers.

The polarizer was popular, but many engineers and geologists had doubts. A German mining engineer went through a metals mine with Philipp Schermuly in 1922 and reported that Schermuly's polarizer indicated ore where later excavations proved there was none.[35]

In 1919, a commission of six scientists (two physicists, two geologists, and two psychologists) went to Schermuly's home in Frankfurt for a demonstration. They learned that the polarizer performed wonderfully when Schermuly could see the target but failed when the targets were hidden from view. The committee published their negative results, but that did not stop the surging popularity of Schermuly's polarizer.[36]

The Great War delayed its spread beyond Germany and Austria-Hungary. When the war ended in 1919, the world opened to Philipp Schermuly. Herman Holtz, a New York City manufacturer of precision instruments, returned from Germany in 1920 with the exclusive American license for the Schermuly Polarizer. Holtz enthused that the polarizer would revolutionize oil and mineral exploration.[37]

Schermuly himself arrived in New York in January 1921. One of his sponsors was John Hays Hammond, one of the most prominent American mining engineers of the era. Investors took Schermuly to iron ore deposits in Michigan and Wisconsin, gas fields in Pennsylvania, and oil fields in Arkansas and Louisiana. He showed to their satisfaction that the polarizer could not only find iron ore, but measure the manganese content as well, detect oil underground and distinguish the quality. The *Engineering and Mining Journal* had seen such hype before and declared: "Wonderful! Colossal! And a never-ending line of suckers slip down the stream."[38]

The editor of *Iron Age* reported that in all field tests, the polarizer never failed. *Scientific American* published his account under the title "The divining rod made respectable," but also included an editorial doubting his claims of success and wondering why they were not reported by German technical journals. The magazine admitted: "On the other hand, nobody ever knows what the science of the future may bring."[39] Schermuly's American advocates formed the Schermuly Polarizer Corporation, headquartered in New York, to market exploration services in North America. No results were publicized.

The Schermuly Polarizer was tried around the world, with generally disappointing results. In 1921 the Geological Survey of the Netherlands tested Schermuly and his polarizer by taking him to newly drilled borings for coal in Limburg. Engineer Schermuly failed the tests.[40]

In 1926 the government of South Africa tried the Schermuly Polarizer to find groundwater in Southwest Africa (now Namibia). The official report briefly noted: "the results were not as good as anticipated."[41]

Another high-profile failure was the oil test at Tuggen, Switzerland, a dry hole drilled to 1,647 meters from 1925 to 1928 on the indication of the Schermuly Polarizer. Swiss geologist Arnold Heim wrote that no geologist would have drilled at that location.[42]

Schermuly prospected for oil in São Paulo state in Brazil in 1927. It was widely reported that his drilling locations resulted in the first two oil discoveries in Brazil, when they were drilled in 1929. But the celebration was premature, and Schermuly's wells were failures.[43]

Philipp Schermuly died in 1929 from malaria contracted while prospecting in North Africa. The polarizer had been popular both in Germany and abroad, but did not long survive Schermuly's demise. He was an accomplished engineer with long experience in mining, and his operation of the polarizer no doubt benefitted from his knowledge of geology and orebodies.[44]

Sons of Schermuly

The commercial success of Schermuly's invention inspired similar inventions by other German engineers. Even as the polarizer faded, another German engineer, Walter Henning, invented a spinning instrument very similar in appearance to the Schermuly Polarizer, though Henning's invention never caught on.[45]

In 1924, periodicals worldwide carried news of the Radio Emanator, invented by Leipzig engineer Dr. W. Pastor. Pastor said that it could detect any kind of mineral deposit within thirty kilometers of the instrument. The instrument appeared similar to the Schermuly Polarizer with the addition of a weak alternating current applied to the sample chamber. A scientific examination in 1928 concluded that the radio emanator was a type of dowsing apparatus and marveled that Pastor had turned uncritical press coverage into three years of lucrative consulting fees.[46]

In 1927, a Munich mining engineer named Englehardt revealed that after four years of work he had perfected a mechanical pendulum that detected oil via "cosmic electricity." The apparatus was not described in detail. He assured doubters that it was all very scientific, based on molecular vibration frequencies of the substance sought.[47]

GERMANY IN WORLD WAR II

When the oil shortage became severe near the end of World War II, German leadership—dismissing the strenuous objections of geologists—resorted to doodlebugs to seek petroleum. SS head Heinrich Himmler ordered that wells be drilled for oil in Schleswig-Holstein based on pseudo-geophysical methods such as the *Geoskop* (Geoscope). Luftwaffe officer Schröder-Stranz invented a supposed death-ray to protect against aerial attacks, but when it proved ineffective as a weapon, Hermann Göring supported his efforts to use his invention to find oil. Schröder-Stranz was still adapting his device for oil-finding when stopped by the Allied invasion in 1945.[48]

THE MANSFIELD AUTOMATIC OIL DETECTOR

The most popular and persistent doodlebug in English-speaking countries was the Mansfield Automatic Water and Oil Detector. Inside were hundreds of wire coils, and on the instrument face was a magnetized needle free to rotate horizontally. The operator noted the amplitude of rapid swings in the needle. The higher the swing amplitudes, the greater the indication of underground water or oil. The Mansfield machine was one of a number of similar machines. These were not mysterious black boxes and did not depend on involuntary muscle movements. Instead, the instruments reacted to small and rapidly changing magnetotelluric currents.

Schmid's Water Finder

The first such instrument was invented in 1903 by Swiss engineer Adolf Schmid, who modified a magnetometer invented in the late 1800s by the French priest Abbé Fortin. Schmid's instrument consisted of iron wire wound 5,000 times around a glass cylinder and placed next to a magnetized needle free to swing horizontally. Magnetic fluctuations induced current in the wire, creating a secondary magnetic field, which in turn caused the needle to rotate. Schmid theorized that magnetic fluctuations were concentrated along water-filled fractures. Schmid's US patent for a "[d]evice for detecting subterranean waters" was granted in 1907.[49]

MANSFIELD'S
PATENT AUTOMATIC
WATER & OIL FINDERS

are used by leading Governments, Railway and Land Companies, Well Boring Engineers, &c.

A Colonial Well Boring Engineer writes :—

"Since purchasing the instrument I have selected 300 sites, most of which I have either bored on with our own plants or had bored on by Sub-contractors to us, and in every instance have been successful."

The prices of the instruments are as follows :

£30 for locating at any depths up to **200 ft.**
£50 ,, ,, ,, **500** ,,
£100 ,, ,, ,, **1,000** ,,
*£175 ,, ,, ,, **3,500** ,,

*Also used for locating Oil.

Delivery at Colonial or Foreign Seaport, £1 extra.

Makers of Well Boring Plant, Pumps, Windmills, Oil Engines and everything connected with water supply and irrigation.

W. Mansfield & Co.,
17, BRUNSWICK ST.,
LIVERPOOL.

Cables :
"MANTLES, LIVERPOOL."

Fig. 22. The Mansfield Automatic Water & Oil Finder was widely advertised from 1908 into the 1930s. (*The Windsor Magazine*, November 1912.)

Fig. 23. The Mansfield and similar instruments detected magnetotelluric currents by the movement of a compass needle. (Photo by the author.)

Mansfield's Patent Automatic Water and Oil Finder

In 1908, water engineer William Mansfield of Liverpool, England, introduced an instrument similar to Schmid's. Mansfield's instrument doubled the number of turns of the wire to 10,000 and used an indicator needle with two north poles, one stronger than the other. The instrument was patented in the United Kingdom, but the US Patent Office denied the patent as too similar to the Schmid device.[50]

The original Mansfield machine was for finding water, but in 1913, the Mansfield company advertised it as finding both water and oil. In 1919 it was available in four models for different nominal depths of investigation, from 200 feet (£30 / $146) to 3,000 feet (£175 / $851). In 1930 there were seven models, with depths of investigation up to 8,500 feet (£500 / $2,432). W. Mansfield and Company was still making and selling its detector in 1933, and the detectors continued in use for decades after.

A Texas oil man tested the machine in 1924 over known oil reservoirs and barren areas, and described it as "most uncanny" in detecting oil. It was tried by a number of American oil operators in the 1920s, from Louisiana to Washington state, but few operators seem to have used it more than once. The Mansfield instrument seems to have persisted in Canada longer than in the US. Canadian oil operators used it widely during the 1920s and early 1930s in British Columbia and Alberta. As in the US, Canadians eventually dropped the Mansfield oil finder. Australian oil prospectors also tried it, with no more success.[51]

Henri Mager's Galvanometric Indicator

Henri Mager was a prominent French dowser in the early twentieth century. For years, Mager used Y-rods and then pendulums to look for water. But when he started oil prospecting in 1912, he left his dowsing rod at home and designed a modified Schmid machine. In 1923, Mager offered his services in using his magnetic detector to find oil.[52]

Mager wrote that the best time to operate was between 9 a.m. and 3 p.m., and that it should only be used on clear dry ground during clear and calm weather, away from trees and buildings. Mager formed a corporation in 1925, *Société d'Etudes et Recherches Pétrolif Scientifiques des Pétroles par Detectores Electroniques Accordes Henri Mager*, to search for oil with the detector. Several wells were drilled on Mager's locations, but none found oil. As did many dowsers, Mager believed that petroleum flows in underground rivers, and he announced that he had used his instrument to trace the source of the Gabian oil field for thirty miles.[53]

Mager chose the location of a test well and drilled in 1928 at Birlenbach in the Bas-Rhin Department. The well hit a promising shallow oil zone, but when the driller tried to explore deeper, the bit stuck and the hole was abandoned. Mager wrote that the drill bit had been stuck "maliciously," but without specifying a motive.[54]

Magnetotellurics

Arthur Ellis, a geologist with the US Geological Survey, reviewed the Schmid, Mansfield, and Mager devices and noted that the premise that telluric currents are associated with water was speculative and unproven. Americans started using magnetotellurics in the late 1950s to search for geothermal energy, and the Russians used it in the 1970s for petroleum exploration.[55] The Mansfield and similar detectors were early induction coil magnetometers. They detected actual physical forces, and so were perhaps the first geophysical devices widely used in the oil industry. They were similar to doodlebugs in one way: they didn't find oil.

TWILIGHT OF THE DOODLEBUGS

THE RISE OF THE GEOPHYSICISTS

Early twentieth century scientists dazzled the public with discoveries such as radioactivity and Hertzian waves, and inventors tried to harness the new science, or at least the jargon, to find oil. And because it is so much easier to invent something that doesn't work, the doodlebugs arrived in the oil fields a decade ahead of real geophysics.

Just as Europe was ahead of the United States in using geology to search for petroleum, Europe also led the way in geophysics. The torsion balance gravity meter came to the US from Hungary in 1922. Downhole geophysical well logging came from France courtesy of the Schlumberger brothers. Seismic refraction equipment, along with German crews to run it, arrived in the US in 1924. Cynics called seismic equipment the German doodlebug, and the seismic crews doodlebuggers—and seismic crews have been called doodlebuggers ever since.[1]

The industry was already crowded with phony geophysical devices, but both gravity and seismic refraction proved very successful in finding oil associated with salt structures on the Gulf Coast. Away from the coast, however, North Texas, Kansas, and Oklahoma were still doodlebug strongholds.[2]

The method that challenged doodlebugs on their home turf was seismic refraction, an American invention. Seismic reflection proved in the late 1920s that it could be profitably applied both along the Gulf Coast and far inland, and it is still by far the dominant geophysical method.[3] An industry writer in 1930 recognized the threat to doodlebugs: "It makes one wonder what will become of the poor doodlebug operator, however, to contemplate the coming of the geophysicist."[4]

Although seismic reflection replaced doodlebugs to some extent, seismic was mostly effective in finding structure-related oil. There is a lot of oil in stratigraphic traps unrelated to structure, and finding that non-structural oil continues to be the appeal of doodlebugs.

What Killed Off the Doodlebugs?

When geophysical methods were introduced in the early 1920s, they initially gave a boost to doodlebugs. Gravity and seismic refraction methods were initially identified as doodlebugs, and their success suggested that other doodlebugs might also be useful. The technical editor of *Petroleum Age* commented in 1924: "The doodlebug has now become a scientific institution."[5]

Universities introduced courses in geophysics—the Colorado School of Mines offered its first course in geophysics in 1927—then degrees in geophysics, and the battle lines were drawn. Geophysicists were keen to separate themselves from the doodlebuggers, and as once-young geologists and geophysicists moved up into management, their views became dominant. Geophysicists remain today the most effective factor in discouraging doodlebugs.

OIL-DOWSING'S DISCOVERY OF THE WEST EDMOND OIL FIELD

Displayed in a prominent glass case at the Oklahoma History Center in Oklahoma City is an elongated hollow metal cylinder hanging by a watch chain, next to an old metal lunch box and some antique bottles of chemicals. It seems an odd and humble assortment of rusted bric-a-brac to be highlighted in a handsome museum exhibit, but these are the tools that oil dowser James M. Young used to find the West Edmond oil field northwest of Oklahoma City.

James M. Young was a farmer from Edmond, Oklahoma, who had taken a fling in oil wildcatting. About 1938, he hired a dowser from Dallas to check some land for oil. Young was impressed with the result, and when he asked for more dowsing tests, the dowser lent him the chemical-filled pendulum.

Fig. 24. The oil-dowsing equipment used by James M. Young to set up Ace Gutowsky's discovery of the West Edmond field. (On display at the Oklahoma History Center, Oklahoma City. Photo by the author.)

Young later bought the instrument from the Texan, but then quit wildcatting and returned to farming.[6]

The pendulum consisted of a metal cylinder filled with a secret chemical and hung from a watch chain. Young held the watch chain in his right hand, and a stopwatch in the left. The pendulum swung back and forth over salt water, but over oil it would swing in a circle, and Young would check the stopwatch to determine depth to the oil. Young said that his doodlebug would not work for just anyone: he was successful because his body had electricity. A geophysicist who watched Young said that the hand holding the watch chain trembled with palsy.[7]

When Young bought farmland, he was guided by his oil-seeking pendulum as much as the quality of the topsoil. Young bought a 320-acre tract west of Edmond that, according to the pendulum, had oil at a depth of 7,000 to 7,100 feet. He had quit wildcatting, so he approached a wildcatter he had worked with before—and most importantly one who would take Young's dowsing pendulum seriously—Ace Gutowsky.

The Wildcatter from Ukraine

Gutowsky's given name was "Assaph," but everyone called him "Ace;" even his tombstone reads "Ace Gutowsky." He came to America in 1902 from Zhytomyr, Ukraine. As he liked to recall, he walked off the ship in Baltimore, fifteen years old, with three cents in his pocket. Back in Ukraine, he had been a Volga German named Gutoff, but somehow he entered America as Gutowsky. He was a strong man and had no trouble finding work. He mined coal in Pennsylvania before moving to Chicago to work in a factory. In gratitude for the opportunity he found, he became an intense American patriot, although he never shook off his thick accent.

According to family lore, one night in 1907, in a bar in Chicago, Gutowsky met Oklahoma farmer Theodore Ladwig, whose description of the fertile farmland of Kingfisher County, Oklahoma, was so compelling that Gutowsky moved to Oklahoma, bought a farm, and the following year wed Ladwig's youngest daughter, Augusta. Gutowsky rotated his land with corn, cotton, and alfalfa, and introduced a strain of cotton new to the area. The Kingfisher *Daily Midget* called him one of the county's progressive farmers.[8] But being a progressive farmer did not make him a profitable farmer, and he had to sell the farm in early 1917.

Oil drilling was booming in Oklahoma, so Gutowsky went into oil. He joined other Kingfisher businessmen to form the Kingfisher Farmers' Oil Company in September 1917 and Flint Rock Oil in January 1918. He was a director of Kingfisher Farmers', and president of Flint Rock. The company and Gutowsky prospered for a while, but by his own account, he went broke in 1920. By 1922, Gutowsky was an independent oil man wildcatting in northern Texas. Gutowsky became president of Chi Oil, and a trustee of Chase Oil in Tulsa.

Gutowsky believed in doodlebugs. In 1925 he announced that he had bought an infallible oil-finding machine in Germany and had used it to select a location for Chi Oil southeast of Kingfisher. The instrument predicted that the well would hit three oil sands. Gutowsky was confident: "The public may laugh at oil-finding machines all it wants to, but they have been proven in Europe, and this one I have has been tested on 400 wells already drilled, and has not failed to indicate yet."[9]

Fig. 25. Wildcatter from Ukraine Assaph "Ace" Gutowsky. (Photo courtesy of Kim Gutowsky van der Wal.)

The doodlebug didn't work very well. Chi Oil gave up at 2,600 feet, stopped drilling and walked away from the well.[10] Gutowsky quit farming and dedicated himself completely to oil drilling. He kept his fondness for doodlebugs, but he didn't enjoy much luck. According to *Time* magazine, he drilled twenty-eight dry holes.

In June 1942, the *Bulletin of the American Association of Petroleum Geologists* reviewed recent developments in Oklahoma in which the author joked that "Methods employed in the search for oil include practically everything except the 'wiggle-stick.'"[11] He no doubt thought that modern geology and geophysics had long since replaced the ancient dowsing rod. But unknown to the writer, the dowsing rod, or wiggle-stick, and the dowsing pendulum were still common in Oklahoma, and one was about to make the largest Oklahoma oil discovery in a decade.

In 1942, Ace Gutowsky leased James Young's farm and surrounding land. He didn't rely only on Young but had sixteen other dowsers and doodlebuggers check out his well location, which was not on Young's land but adjacent to it. Thirteen of the dowsers and doodlebuggers agreed with Young that there was oil under Gutowsky's location.[12]

Gutowsky did not ignore geology. The Gypsy Oil Company (part of Gulf Oil) had drilled a deep dry hole on the southeast edge of the structure, and some other major oil companies had held leases in the area, though none considered it quite good enough to drill. Gutowsky had maps of the subsurface that showed what he believed was an oil-bearing structure at the Wilcox Sand.[13] Gutowsky and his partner, real estate agent B. D. Bourland, looked for someone to pay for the drilling.

Ace Gutowsky had a very difficult time finding investors, but finally convinced some Chicago businessmen to pay for the wildcat well. He started drilling in January 1943. The purpose of the well was to test the Wilcox Sand, but in March 1943 the well went through the Wilcox Sand at 7,350 feet and found it to be wet.

On the way down to the Wilcox, however, the well had passed through the Hunton Limestone, and the bits of crushed Hunton Limestone that circulated up to the surface had traces of oil, a "show," so Gutowsky took a chance and installed casing in the well to test the Hunton. Oil flowed to the surface at 522 barrels of oil per day, and Gutowsky was in business.[14]

Some thought that Gutowsky's discovery might be a one-well field, but another company drilled a well a couple thousand feet away and also found oil. More wells followed and the field expanded. A year after the discovery, the field had fifty wells and had found its limit only along part of the southwest side. Companies realized that the new field was very large and scrambled to buy leases ahead of the drill rigs. By late 1944, 120 drilling rigs were boring away in the field, the largest concentration of rotary drilling rigs in the world.

Within two years of Gutowsky's discovery, the field had more than 500 oil wells, was seventeen miles long, four miles wide, and extended into four counties. Gutowsky and his wigglestick men had discovered the giant West Edmond field, the largest oil field found in Oklahoma in ten years. The discovery changed Gutowsky from a hopeful wildcatter to an oil millionaire. He always gave credit for the discovery to the doodlebugs and dowsers, especially to James M. Young.[15]

For the next several years, Gutowsky kept three dowsers and doodlebuggers on the payroll, but found that they could not eliminate the risk. Six months after his West Edmond discovery, Gutowsky drilled a dry hole in Pottawatomie County, at a location where doodlebugger A. T. Campbell had predicted a thousand barrels per day. Gutowsky lamented: "Some of the bugmen are all wet."[16]

James M. Young profited from the oil royalties, one-eighth of the production on his 320 acres. Although the West Edmond discovery made him famous, it is not known if he ever went oil-dowsing again. He died in 1957.

The West Edmond oil field is probably the largest oil field ever found by dowsing.

A Pair of Aces

Gutowsky's son Leroy—born in Oklahoma, despite all the mistaken sources that confuse him with his father and say he was born in Russia—also earned the nickname "Ace" when he played eight years as a lineman in the National Football League. The high point of his career was playing as substitute quarterback for the Portsmouth, Ohio, Spartans (before the franchise became the Detroit Lions) in the 1932 NFL championship game when Dutch Clark, the Spartans' All-Pro quarterback, left at the end of the regular season for his main job, coaching basketball at Colorado College, and the college would not grant him leave for the NFL's first championship game.

The teams played for the championship in Chicago on a sixty-yard field in an indoor ice rink because of bad weather. Despite his unfamiliar position, Ace Gutowsky quarterbacked the Spartans through three quarters of a scoreless tie against the Chicago Bears, but finally threw the interception that set up Bronko Nagurski's controversial touchdown pass to Red Grange, putting the Bears ahead. With four minutes left, Gutowsky was returning a kickoff down the sideline when Bears coach George Halas stuck out his foot and tripped him.

But the officials had not seen the trip, and the Bears carried away the championship. Like the other Spartan players, Leroy Gutowsky earned an extra $175 for playing in the NFL championship.[17] He is a perennial nominee for the Pro Football Hall of Fame.

The Wildcatter in Retirement
On February 6, 1945, the elder Ace Gutowsky died of a heart attack in Hollywood, California—or so said all three attending doctors who could find no vital signs. But Ace was tougher than anyone thought, and he came back from the dead. He revived in the hospital and returned to Oklahoma. He explained, "I had unfinished business."

Gutowsky used the profits from West Edmond to go back to wildcatting, with mixed results. He made some good discoveries and drilled some dry holes. In 1946 he bought orchards and farmland near McAllen, Texas, and in 1948 he bought a home near his orchards and retired from the oil business. He explained, "Oil wells give out after a while, but citrus keeps on growing." He died for the second and final time in 1951, at age sixty-four.[18]

A. T. Campbell's Doodlebugs
In May 1941, oil man A. T. Campbell ceremoniously threw his untrue doodlebug into the Canadian River at Shawnee, Oklahoma, and said that he would never again trust a doodlebug. His machine had betrayed him by leading him to drill a dry hole at Lake Tecumseh. But Campbell could not resist the siren song of the doodlebug, and by April 1942, he had another doodlebug. This one, he swore, would certainly lead to oil.

By 1942, dowsing and doodlebugging were on the decline. Yet Campbell estimated that there were several hundred independent operators of oil-finding instruments in Oklahoma alone—evidently the last stronghold of dowsing and doodlebugging. He thought that those dowsers and doodlebuggers should organize to promote their potential contribution to oil exploration.

The last time Oklahoma dowsers and doodlebuggers had tried to meet, it had turned into a public relations disaster. The National Doodlebuggers Convention was held in July 1939 at Ghost Mound near Weatherford, Oklahoma. The theme was to be a search for a legendary quarter million dollars in gold supposedly buried by returning California 49ers just before being killed by Indians. But the organization president didn't show, and before the doodlebugs could warm up or the dowsing rods swing into action, spirit mediums took over. Half a dozen mediums wandered in different directions, and each proclaimed a different spot to dig. The digging crew was unsure of which spirit to follow until they spotted some wild plums, and the convention broke up to eat wild plums. Newspapers nationwide described the event as a farce.

Campbell tried to form dowsers and doodlebuggers into the Instrument Exploratory Association. As self-appointed president, he called its first convention in Tulsa for three days in May 1942. The purpose was to publicize doodlebugs as serious instruments and to convince oil companies to hire them. But dowsers and doodlebuggers proved too independent, and only about ten showed up, outnumbered by reporters. The first convention was also the last.

Despite Gutowsky's West Edmond discovery, and despite the efforts of A. T. Campbell, the days of doodlebugs were running out. The West Edmond discovery was celebrated in newspapers and magazines, but it failed to stop the relentless advance of geologists and geophysicists and consequent decline of doodlebugging and oil-dowsing.

THE DOODLEBUG FROM OUTER SPACE

It was 1947 and Denver oil man Silas Newton needed money badly. He needed to convince investors that they would make a lot of money by funding his oil drilling, but he had an unfortunate habit of drilling dry holes. To woo investors, he got a doodlebug, or at least told investors that he had a doodlebug; he said he had several doodlebugs. He wrote that his instrument detected microwaves supposedly broadcast by underground oil. He told another investor that he had an oil-finding cosmic ray detector built for his exclusive use by a Nobel Prize–winning physicist. But he needed something more attention-grabbing, so he and another con man, Leo GeBauer, made up a story about an oil detector they took from a crashed flying saucer—now that would really command attention.

The financially squeezed Newton of 1947 might have felt a twinge of nostalgia for the 1920s, when he was chauffeured in a twelve-cylinder Cadillac to and from his Park Avenue, Manhattan residence. He had been a multimillionaire, back when that meant something. But it fell apart in the financial crash of 1929, and Newton had fought to regain his wealth since.

Silas Mason Newton was a go-getter of unsurpassed energy and ambition, a star athlete in college, and a top amateur golfer. He moved to New York City in 1925, where he promoted oil company shares. Friends called him "Cowboy," but financial journalist A. Newton Plummer called him "an oily promoter."[19]

Newton's financial gymnastics peaked in early 1929 when he merged Indiana Oil and Gas with independent oil man John Choat and Grayburg Oil. Newton's future seemed limitless as the president of a rapidly growing oil company. Also in 1929, Newton married Nan O'Reilly, a sportswriter for the New York *Journal*. It was his second marriage. Although she lived with Newton and ten servants, Nan feared that her editors would fire a married female employee, so they kept their marriage secret. Newton's chauffeured Cadillac drove her to work each morning but dropped her off around the corner from the *Journal*, out of sight of her coworkers. The secret marriage was finally revealed, but, perhaps because of the national publicity, the *Journal* let Nan O'Reilly Newton keep her job.[20]

It seemed that nothing could stop Silas Newton. But instead of the beginning, this would be the peak of Silas Newton's career.

The Grayburg Oil merger unraveled after the great stock market crash of October 1929. A stockbroker sued Newton for a $125,000 commission, Grayburg Oil won a $300,000 judgement against Indiana Oil and Gas, and against Newton personally, and won the right to sell off $400,000 of Indiana Oil and Gas bonds. The failed merger financially crippled both companies.

Newspapers reported Newton to be worth forty million dollars in 1930, but his wealth evaporated. He did not stop living the good life—he just stopped paying for it. From 1932 to 1935, merchants obtained multiple court judgements against the Newtons.[21] Newton tried to repair his fortune by promoting startup oil and gas companies. In the 1920s, it had been easy: investors clamored to buy shares, and printing share certificates was almost like printing money. In the 1930s, investment money was scarce, and most ominously for Newton, state and federal governments passed laws regulating stock sales and went after high-flying financiers who did not comply.

Newton repeatedly landed in legal trouble for selling oil shares through misrepresentation. In 1931, the New York Bureau of Securities arrested Newton on the fraud complaint of a man who paid Newton $25,000 for Indiana Southwestern stock. Lured by Newton's salesmanship, the man had invested all he had, plus $18,000 borrowed on his home, before discovering that the shares were worthless. Newton apparently settled the matter out of court.[22] In 1933, Newton was called in once more by state securities regulators on a complaint that Newton had sold shares in Indiana Oil and Gas for $25,000 by misrepresentation.

Newton was arrested in 1934, along with five people with the Benjamin Baker stock brokerage. All were charged with grand larceny for selling worthless stock and for issuing prospectuses with false statements—and all were connected to Silas Newton's oil and gas companies, including Indiana Eastern, which the assistant district attorney called "absolutely worthless." Luckily for Newton, the main target was stockbroker Benjamin Baker. Baker pled guilty, and as part of the plea deal, the state dropped charges against other defendants, including Silas Newton.

In 1935, Walter Winchell, the dean of New York gossip columnists, wrote in a column of one-liners that also featured the latest doings of aviator Charles Lindbergh and fan dancer Sally Rand: "Silas Newton, the millionaire who went bust in the crash, has struck oil again."[23] But Newton had not come back. He kept trying to prop up insolvent companies by splitting, merging, and organizing new companies to rescue the old, but the whole lot finally collapsed.

Nan O'Reilly Newton died in 1937. The following year Silas Newton moved to Denver, occupying both a house in the exclusive country club neighborhood and rooms at Denver's premier hotel, the Brown Palace.[24] He left New York, a metropolis of seven million, whose cosmopolitan sophistication he enjoyed, and the financial center he had relied on for his lifestyle. Denver was then a backwater of the oil industry, a city of three hundred thousand at the western edge of the plains.

Newton was apparently lured west by the Oriental Refining Company, an organization created to take over and expand a refinery north of Denver. Oriental no doubt saw Newton's experience as just what they needed, and Newton may have seen Oriental Refining as the chance to lead a small, but growing, vertically integrated oil company, as he had tried to do with Grayburg Oil.

Newton became president of Oriental Refining in 1939 but left the company within a year to go into business for himself. He became the operator of seven already-producing oil wells in Wyoming and New Mexico. He also operated a few wells in Kansas, but the handful of old wells could not satisfy either Newton's ambition or his expensive tastes. So Silas Newton went wildcatting. He drilled four dry holes in California in 1939 and 1941.

Newton later bragged that he had "rediscovered" the Rangely oil field in Colorado. He told gullible Frank Scully that he had rediscovered it using his secret oil-finding device. However, the Rangely field, in remote northwest Colorado, could not have been "rediscovered" because it was never lost. The field produced small volumes of oil from numerous shallow wells in the Mancos Shale since its discovery in 1902. To oil men, Newton's claim of rediscovery might imply that he was the one who found the deep oil pool in the Weber Sandstone—which would certainly give bragging rights, if true. It was the deep discovery that brought little Rangely into the ranks of "giant" oil fields. Rangely has produced more than one billion barrels of oil.

Contrary to his bragging, Newton had nothing to do with the deep discovery at Rangely. The California Company (the corporate ancestor of Chevron) discovered the deep oil at Rangely in 1933 by drilling what was then the second deepest well in Colorado. The borehole intersected a massive oil deposit: a six-hundred-foot thickness of oil-saturated Weber Sandstone, and a full-page headline in the *Oil & Gas Journal* shouted, "Rangely Dome Discovery of Major Importance." But at the depression oil prices of 1933, the oil flow from the deep well at Rangely was not enough to pay the cost of deep drilling and transportation from the remote location. No more deep wells were drilled at Rangely for the next ten years.[25]

The prodigious amounts of oil required by World War II increased the oil price, making the previously uneconomic deep Rangely oil a bonanza. Newton visited Rangely in 1942 and saw the future: a landscape dotted with oil rigs digging down to the oil-soaked Weber Sandstone. He saw his own fortune revived and larger than ever, except he was about ten years too late. Almost all of the Rangely anticline was owned by three majors: the California Company, the Texas Company (Texaco), and Stanolind (Amoco). They knew they were sitting on a giant oil field and were not interested in selling.

Newton found his chance when he met John Bockhold, a small-time oil man from Kansas who had leased 1,759 acres along the south edge of the Rangely anticline but needed to sell his lease to pay debts. Nobody yet knew the exact extent of the Weber oil pay, and with a bit of luck, the deep Rangely field could extend into the northern part of the Bockhold lease—with a lot of luck, perhaps a sizeable part of the lease. Newton agreed to buy the leases for $250,000, but arranged to pay only $4,769 cash, another $5,000 after one year, and the remainder to be paid from future production.[26] Bockhold and his creditors didn't know that once Newton had the lease in his name, he would not pay them another cent.

Newton's first two wells ran into drilling problems and failed to reach the Weber. He sold part of the Bockhold lease to oil operators who paid the entire cost of drilling a well in exchange for 50 percent ownership. Here was Newton at his deal-making best. He paid nearly nothing for the lease, paid nothing to have it drilled, and came out of the deal owning partial interests in four oil wells—but it forced him to give up half of the tract. He still needed money, so he sold a 12.5 percent royalty in the Newton-Cobb-Stringer acreage to oil man Paul Beamer, and he sold royalties in another tract to a Denver physician and his wife, Wenzel and Mable Friesch.

His Newton Cobb-Stringer Government #1 drilled into the Weber Sand in June 1946. Despite Newton's announcement that the well had oil-saturated Weber Sand, which would have meant an oil producer (Newton almost always described his wells as having oil-saturated sand), he could coax only water from the well and it was a dry hole. The Associated Press picked up the story, and newspapers across the country reported that after seventy-one successful oil wells to the Weber at Rangely without a single failure, Silas Newton had drilled the first dry hole in the Weber Sand. Newton's dry hole condemned almost all of his oil lease.

While he was pouring money into Rangely, Newton was also busy with far-flung wildcats—all dry holes. In 1946 and 1947, Newton drilled five dry holes in New Mexico, Colorado, Kansas, and Wyoming. Newton's unpaid field contractors had the surface equipment seized at Newton's New Mexico wells, and because Newton was no longer producing oil, the landowner cancelled his oil lease. Newton likewise left Kansas without paying the driller and an oil field supply company, so they sued.[27]

Newton tried to exaggerate the value of his Rangely lease when he drilled the Government #14 a mile south of the Weber pool. He knew, or should have known, that it would be a dry hole. In March 1947, he announced that the well had oil-saturated Dakota Sandstone 600 feet shallower than shown on the US Geological Survey map. Had this been true, it would have proven that another anticline lay to the south, another Rangely field right in the middle of Newton's lease. Most industry people recognized that what Newton called the Dakota was actually the shallower Frontier Sandstone. After Newton's loud self-congratulation, the Government #14 made no oil, only a little gas, and was shut down after three months. Later wells proved that Newton's report was false, and that the well was drilled to the shallower horizon.

Stanolind operated the four associated wells, of which Newton owned 10 percent, but never paid his share of expenses. Stanolind sued, and the court awarded them Newton's share of oil production, until Stanolind recouped the $23,120 debt.

Newton never paid a cent of the royalty he owed to the Bockhold creditors and refused to even give an accounting—so they sued Newton in April 1950. Newton obfuscated and delayed and claimed that their lawsuit was barred by laches and the statute of limitations. The court would have none of it and ordered Newton to pay $67,317. Newton did not have the money.

Newton likewise never paid anything to Mable and Wenzel Friesch, so they sued in February 1951. Newton denied that he owed them anything, but the court ordered Newton to pay the Friesches $31,862. Newton didn't have the money, so the court ordered Phillips Petroleum, which operated a well in which Newton had partial ownership, to pay Newton's share directly to his creditors. Phillips complied but wrote to the judge that it might be impossible to satisfy everyone to whom Newton owed money.

Newton had, by means fair and foul, sucked all the money possible out of the Rangely lease, and he and his oil company were being dunned by contractors and royalty owners. Creditors filed eleven lawsuits in the Denver court against Newton from February 1948 to April 1952, seeking $123,200 for unpaid royalties, rent, field equipment and services, wages, and payments on notes of indebtedness. Newton needed money badly.[28] So he made up some doodlebugs.

Magical Microwaves and Other Doodlebugs

In the *1947–1948 Rocky Mountain Petroleum Yearbook*, Newton wrote an article claiming (falsely) that oil deposits emit microwaves that can be detected at the surface. This startling new discovery would eliminate dry holes. Why microwaves? Microwaves were in the news as the latest in long-distance communication. The suckers couldn't understand the scientific jargon—and good thing, because Newton didn't understand it either—but they had read something about those amazing microwaves and that left them feeling a little less befuddled.

Oil men knew that microwave-broadcasting oil was nonsense, but Newton's quirky articles in the *Oil Reporter* and the *Rocky Mountain Petroleum Yearbook* gave him legitimacy outside the industry, even as they discredited him to industry people. He published another false write-up about finding oil by microwaves in a 1949 Yale alumni publication.[29] In the past, Newton's lies about stocks or oil prospects were verbal, but this time he lied in print, and there was no way to deny or justify it. He was an undisputed con man.

Newton knew that it was easier to scam money from people if he appeared to be so wealthy that he didn't need their money. Even as his debts piled up, he wore expensive suits, sported a healthy tan, and always drove a new Cadillac.

Newton told optometrist Alfred Kleyhauer that Nobel Prize–winning physicist Robert Millikan had developed a cosmic ray detector for Newton's exclusive use in finding oil. Newton said that he had used the detector to rediscover the Rangely field, and that he was so certain his machine would find oil that he gave Kleyhauer a money-back guarantee—literally too good to be true. In 1946 the optometrist gave Newton $15,000 to use cosmic rays to locate more oil wells in Colorado and Kansas, and for a royalty on Newton's Rangely oil wells. Once Newton had his money, he delayed and made up excuses, until after two years of run-around, his victim became suspicious and wrote to Robert Millikan at Cal Tech University. Millikan wrote back that he knew nothing of either Silas Newton or of the cosmic ray detector. Kleyhauer could not find a lawyer willing to take quick action, and the three-year statute of limitations for fraud expired.[30]

Leo GeBauer: "Dr. Gee"

Silas Newton found a partner in Leo Arnold Julius GeBauer, the owner of a radio parts store in Phoenix who had branched out into making oil-finding doodlebugs.

GeBauer grew up in western Nebraska and bounced between jobs in sales and retail. He was investigated by the FBI in 1941 for making such statements as: "It would be God's blessing if we had two men in the United States to run this country like Hitler." That and his hobby as an amateur radio operator raised suspicions that he was a spy, but an FBI agent cautioned: "Some of the rumors have to be discounted because he is intensely disliked by his neighbors." In May 1942, GeBauer took a civilian job with the Army Signal Corps as a junior radio operator, but the Army fired him when they learned of his Nazi sympathies. In 1942 he was arrested for violating the federal housing act. He pled no contest and was given a one-year suspended sentence.

Despite being on a US Army list of security risks "considered unfit for defense work," from 1944 to 1947, GeBauer worked for the AiResearch Company in Phoenix, Arizona. An FBI memo described his job as "mainly in charge of maintenance of equipment." He started a radio parts store in Phoenix in 1949 and opened a branch in Denver in 1952.[31]

He also tinkered with electronics to detect oil and groundwater, and by 1949 he had developed his doodlebug and worked for oil drillers in Arizona and Colorado. GeBauer claimed degrees from five universities, but the FBI found no record of him at any of the institutions. GeBauer also said that he had led a top-secret team of hundreds of scientists studying the northern lights.

In 1949, GeBauer wrangled an introduction to Denver industrialist Herman Flader, and showed Flader a device he said could find oil and water in the ground. Silas Newton also introduced himself to Flader and demonstrated his own oil-finding gizmo. Flader played Newton and GeBauer against one other, letting them criticize each other's oil-finding machines. Although Newton and GeBauer pretended that they had never met, they had known one another since at least the previous year. Newton told Flader that GeBauer was one of the nation's top scientists, and that GeBauer's gizmo was just the thing to find oil on tracts that Newton had under lease.

With supposed rival Silas Newton vouching for GeBauer's oil-finding machine, Herman Flader bought two of the gizmos from GeBauer for $31,000. They were so expensive, GeBauer told him, because they had plutonium-tipped antennae, and a tube of plutonium inside. But Flader was never to peek inside, because if opened, they would explode. The instruments were actually Army radio tuners being sold at surplus stores for a few dollars.[32] Herman Flader gave Newton and GeBauer more money to drill wells in Colorado, California, and Wyoming. He estimated that he spent $250,000 on Newton and GeBauer and received nothing in return.

Flader funded the Paul Beamer #1, drilled in the Mojave Desert in 1948, known as the wildest of wildcat wells. In early 1949 Newton announced that the well was down to 1,200 feet and drilling through oil-bearing sands. Despite the "oil-bearing sands," the well was a dry hole. In November 1949 Newton drilled another Mojave Desert dry hole, about a mile from the first.[33]

Newton told Flader that there was deeper oil pay at the Dutton Creek field in Wyoming. Flader and GeBauer went to the field, and GeBauer's doodlebug confirmed that the Sundance Sandstone was filled with oil, just as Newton said. Flader reentered the exhausted Eades #27 and drilled it down to the promised oil sand, but the well ran into problems and was plugged. Newton argued that the well should have yielded two thousand barrels per day from the Sundance

Sandstone but was wrecked because Flader had installed corroded casing, which collapsed. Flader countered that even though the casing had collapsed, the well had flowed water, not oil, from the Sundance. Despite a number of later tries nearby, no oil has ever been produced from the Sundance on the Dutton Creek anticline.

The Show Business Columnist

In the 1940s, Newton established residences in Los Angeles: a private home in Hollywood, and rooms at the Hollywood Knickerbocker Hotel. Newton needed investors for his drilling projects, and show-business people in particular were attracted to the tax advantages. Newton cultivated a friendship with Frank Scully, who wrote a column for *Variety*, dishing up inside tidbits with a slangy Damon Runyon style. The friendship was golden for Newton because Scully knew everyone in Hollywood. Scully's daughter Moreen described Newton as "Kind of flashy with big bow ties. He wore a lot of turquoise . . . always drove a Cadillac."[34]

Scully described Newton as "one of the great geophysicists of the oil industry," who had "rediscovered the Rangely oil field" using his own secret geophysical instruments. This was nonsense, but Scully believed whatever Newton told him.

Scully introduced Newton to redheaded, aspiring fashion designer Sharon Chillison, who had moved to Los Angeles from Detroit after her divorce. In January 1950, Newton and Chillison were married by a justice of the peace in Tucson, Arizona. She was thirty-three. He was sixty-two (although he put fifty-five on the marriage license). Seven months later, she gave birth to a son, Howard Newton.

Newton introduced Scully to Leo GeBauer, and GeBauer further imposed on Scully's credulity. Scully lapped it up and described GeBauer, alias "Dr. Gee," as "the top magnetic research specialist in the United States."[35] According to Scully, Dr. Gee headed a top-secret wartime research program of 1,700 scientists, which developed a magnetic submarine-detector.

The Doodlebug from Outer Space

By 1949, both Newton and GeBauer had been peddling their doodlebugs with some success but needed to attract more investors. What sort of technology was so advanced that no one could doubt that it would find oil?

According to a memoir in Newton's handwriting that surfaced many years later, Newton and GeBauer were driving through the desert in August 1949 when they conceived of the ultimate doodlebug: an oil-finding machine that GeBauer had taken from a downed flying saucer. Flying saucers were a new sensation, and this seemed like a brilliant way to make their doodlebug stand out from the competition. GeBauer had just returned from a trip promoting his doodlebug to oil operators in northwest New Mexico, near the town of Aztec, so the men chose that area as where a saucer had landed.[36] Aztec was—still is—a pleasant cottonwood-shaded ranching town where the Animas River flows among juniper and pinyon pine–covered mesas.

GeBauer's and Newton's tale of a UFO landing in New Mexico was not entirely original. In June 1947, a secret Project Mogul balloon with a payload to detect far-off atomic explosions had blown away from the Alamogordo Army Air Base and landed near Roswell, New Mexico. A rancher told the county sheriff that he had found some odd debris, perhaps from one of those flying discs. The sheriff called the Roswell Army Air Field, which sent soldiers to gather the debris. The debris was unlike any they had seen, and the base issued a press release that they had recovered wreckage of a "flying disc." But the Project Mogul scientists realized that the flying disc was their lost balloon; they took the debris and put out the cover story that it was just a weather balloon, which some people believed—others did not.[37]

Newton and GeBauer improved the Roswell story by adding alien corpses. Their Aztec saucer had not actually crashed but landed on autopilot after something had punched a hole in a glass porthole and the crew died. By an amazing coincidence, the dimensions of the saucer were in exact feet and inches (take *that*, metric system!). The crew (probably from Venus, they said, because at forty inches tall they were too short to be from Mars—this sounds like interplanetary dating advice) had died, but scientists salvaged new technology from the saucer. The government was keeping the whole thing hushed up, of course, but GeBauer helped himself to some equipment from the saucer and he and Newton were using it to find oil.

How to get the word out? Frank Scully wrote a weekly column in a national periodical. Scully trusted Silas Newton as a friend and he didn't know enough science to see through such obvious posers as Newton and GeBauer. Newton and GeBauer wowed Scully with their wild tale of a saucer landing, and Scully wrote the crashed saucer story in *Variety* on October 12, 1949. Newton was soon pitching his doodlebug from outer space to Hollywood investors. In November 1949, actor Bruce Cabot told the FBI that at the Lakeside Country Club, Silas Newton had been talking up a "magnetic radio" he had taken from a crashed saucer and was using it to find oil wells with out-of-this-world ease.[38]

Newton's friend George Koehler helped spread the story. Koehler, the advertising manager for a Denver radio station, lived in Newton's house and drove a Cadillac registered to Newton. In Koehler's version, he had talked his way onto a top-secret radar station where he saw two downed flying saucers. The base employees even let him take saucer parts as souvenirs, which he showed as proof.[39]

A student in a Basic Science class at the University of Denver heard Koehler talk about saucers and suggested that the instructor, Francis Broman, invite Koehler to lecture on flying saucers. The class was enthusiastic. Koehler declined but agreed to ask a scientist friend who was an expert. Newton accepted on the condition that he would not reveal his name. Broman agreed and told his class that they should consider the lecture as an exercise in critical thinking. Word spread, and at half past noon on March 8, 1950, Newton walked into an overflowing classroom, where he proceeded to detail his wild tale of crashing saucers and recovered alien bodies.

The reaction was sensational. The talk was written up in the next day's Denver *Post*. The lecturer was identified only as "Mr. X," but one of the students in the class had caddied for Newton and, on March 17, the Denver newspapers identified the mystery speaker as Silas Newton.[40]

Frank Scully contracted with a publisher to write a book on flying saucers. He pounded his typewriter furiously, and in seventy-two days produced *Behind the Flying Saucers*.[41]

To help Scully, Silas Newton recorded hours of audio, spouting an eccentric blend of science and nonsense on astronomy, magnetism, and flying saucers, much of the information supposedly from the top magnetic expert in the United States. In some recordings, Newton read from textbooks, leading Scully to unknowingly plagiarize.[42] The tapes reveal that Newton was the main source of scientific error in the book, as Newton and Scully created a powerful dynamic of the ignorant leading the gullible. *Behind the Flying Saucers* remains a test of scientific literacy: if you can read the book without realizing that it is a jumble of pseudoscientific stupidities, then you fail the test.

The publisher rushed *Behind the Flying Saucers* into bookstores in September 1950. As a hardcover book by reputable publisher Henry Holt, it was widely reviewed. Reviewers recognized the gross scientific errors in the book, and its credibility unraveled. Before printing a review, the *Saturday Review* asked Scully if the book were not an out-of-season April Fool's joke; Scully assured the magazine that he was serious. The *Saturday Review* noticed errors of grade-school-level science and "blatant inconsistencies and misstatements of facts." *Time* magazine turned its Buck Rogers blaster ray on the book and declared, "Scully's science ranks below the comic books." The reviewer for the *Washington Post* wrote: "No one with even a primary knowledge of the various sciences could have read proof on the book."[43]

Scully did not suffer criticism in silence. Although he had an amputated leg, and spent most of his life in chronic pain, he was a scrapper always ready to fight. The Catholic Church had made him a Knight of St. Gregory and a Knight of St. Elizabeth, and Scully took his knighthood seriously. He filled the two volumes of his autobiography, titled *This Gay Knight* and *In Armour Bright*, with stories of him battling injustices large and small. But there was no space in Scully's crusading spirit for respectful differences of opinion. Anyone who contradicted his friend Newton must be stupid or part of the sinister government cover-up. Scully toured New York, Chicago, and Denver, giving radio and television interviews blasting his critics as "Pentagonian stooges."[44] The book spent sixteen weeks on the *New York Times* best-seller list, peaking at number four, and was translated into twelve foreign languages.

Freelance writer John P. Cahn read *Behind the Flying Saucers*, noticed the scientific errors, and decided that UFOs deserved a more serious treatment. He got the go-ahead from his old editors at the San Francisco *Chronicle* to write a follow-up series for that paper and contacted Scully in February 1951. Cahn described Scully as an "absolutely super guy," but the liking was not mutual. This was Scully's big story, and he had planned follow-up best-sellers. Why help this youngster who had the nerve to think that he could improve on the work of veteran journalist Frank Scully? Scully was proud of *Behind the Flying Saucers* and bristled at criticism. He was the fearless reporter fighting to print the story of the century. But Cahn showed up at Scully's house one evening when Newton was there, and Scully introduced the two. Newton agreed to meet Cahn the following week in San Francisco.[45]

Cahn met Silas Newton in the restaurant at the Palace Hotel, where Newton was evidently well known. Newton relished the spotlight, expanded on the saucer tale, and hinted about his top-secret government work. He dramatically drew a handkerchief from his coat pocket: wrapped in the handkerchief were four small metal disks, which Newton assured Cahn were from the crashed saucer. Newton said that tests had proven that the disks could not be melted and were made of a metal unknown to earthly science.

Scully described Newton as one of the nation's foremost oil executives, but when Cahn asked prominent oil men in Los Angeles, none had even heard of Newton. Cahn visited the Newton Oil Company headquarters in Denver and discovered that it was just two rooms and a reception area with only a couple of employees. Clearly, Newton was a small player. Newton told Cahn that he owned the Oriental Refining Company, but when Cahn phoned the refinery and asked about Newton, the quick reply was, "What's the matter, does he owe some money?"[46]

Cahn was getting nowhere. So far, he had a book full of scientific errors written by a Hollywood columnist, and some unbelievable claims by a minor oil-business hustler with a spotty reputation. The only solid evidence was the metal disks that Newton had told him were unknown to earthly science. Cahn had to get one of the disks; if tests showed that the metal was not of earth then he would have the story of the century.

Cahn asked Newton to submit one of the disks to an independent laboratory, but Newton said that he would not cooperate further unless Cahn paid him; after all, Scully was making a lot of money from his book and Newton wanted some for himself. Cahn told Newton that his publisher would pay $10,000 down, with an additional $25,000 once Newton had delivered the full story of the saucers. Newton played hard to get, but he needed money badly. Cahn had made some substitute disks which he carried in his pocket; he even hired a magician to train him to palm small objects. Cahn made the switch and pocketed the outer-space mystery metal. Tests showed that the disk was just ordinary industrial grade 2S aluminum, commonly used for pots and pans.

Cahn tracked down "Dr. Gee" GeBauer behind the counter at his radio and TV parts store, where he nervously denied being Scully's Dr. Gee.[47] Cahn returned to Scully, who—according to Cahn, but later denied by Scully—admitted that GeBauer was Dr. Gee. Cahn told Scully that Newton and GeBauer were not top scientists, and that the saucer-crash story was a scam.

Scully was fiercely loyal to his friends, and he and his wife Alice were incensed that Cahn would smear Silas Newton.

When Scully and Newton learned that Cahn had submitted the story to *True* magazine they threatened a libel suit, but *True* published Cahn's exposé, "The Flying Saucers and the Mysterious Little Men," in September 1952. Cahn had exposed Newton's lies but didn't understand the motive until a friend mailed him a classified ad clipped from a newspaper:

> URGENT NOTICE—All persons having dealings with SILAS M. NEWTON, Denver, Colorado, New York, Illinois, Wyoming, Calif., Ariz., etc. relative to oil investments, "Cosmic Rays," and/or "Flying Saucers," kindly contact Box M5743 by letter or wire. THIS IS MOST URGENT.[48]

The ad led Cahn to Denver optometrist Alfred Kleyhauer, whom Newton had fleeced of $15,000. The optometrist wanted to warn other victims, and so put small advertisements in newspapers.[49] Kleyhauer told Cahn that Newton and GeBauer had been getting money from Herman Flader. When Cahn spoke to him, Flader realized that he had been bilked and complained to state and federal authorities. The Denver district attorney obtained warrants for Newton and GeBauer.

In October 1952, just ahead of the three-year statute of limitations, FBI agents arrested Silas Newton and Leo GeBauer. Supposed multimillionaire Newton could not pay the $5,000 bond, but his son paid $4,000, and his wife raised the remainder by selling jewelry. The arrests triggered an avalanche of lawsuits by Newton's investors and creditors. One such person was Herman Corsun, a delicatessen owner in Phoenix who, under the spell of GeBauer's oil-finder-from-a-crashed-saucer story, had paid $3,350 for some Wyoming oil properties that GeBauer didn't even own. GeBauer settled out of court for $2,500 and two TV sets.[50]

The federal grand jury refused to indict Newton and GeBauer in February and March 1953.[51] But the state charges for fraud remained, and after delays due to GeBauer's ill health, the trial opened in November 1953.

Doodlebugs on Trial

The prosecution introduced into evidence the two top-secret oil-finding machines that Leo GeBauer sold to Herman Flader for $31,000. They were table-top metal boxes one or two feet in each dimension with a pair of antennae. GeBauer had told him that the oil-finding machines worked on the same principle used by flying saucers. The prosecution brought in two Army radio tuners bought for $4.50 each at a surplus store, nearly identical to the machines identified by Flader as GeBauer's doodlebugs. The only differences were that Flader's machines had dials, lights, and a battery to make the lights flash.[52]

The defense concluded with Newton and GeBauer. Newton went on the offensive, swearing that he had not cheated Flader but that Flader owed him $50,000. Newton was cool and confident under questioning. It was up to Leo GeBauer to end on a high note.

GeBauer's testimony was a disaster for which Newton later blamed his guilty verdict. GeBauer said that his machine detected oil by new forces unknown to science. He gave the jury a chalkboard lecture, but on cross-examination was unable to explain a mathematical formula that he had written. GeBauer told the jury that his machines worked on the same principle as the northern lights, on which he was a top expert. But when the prosecuting attorney asked him about the aurora borealis, GeBauer admitted that he couldn't even spell it.[53] GeBauer also testified that he had twice debated physics with Albert Einstein, but on cross-examination admitted that the debates were actually public lectures by Einstein, after which GeBauer posed questions from the audience.

GeBauer denied ever seeing the two machines made from Army surplus radio tuners that Herman Flader had testified that GeBauer had sold him. GeBauer submitted into evidence two oil-finding machines. The first was a Mansfield Automatic Water and Oil Finder. GeBauer's second doodlebug was of his own making: a steel box the size of a small table, with a light, an altimeter, and some other dials mounted on the face. Using the two oil finders, GeBauer said that he found oil with an accuracy of 86 percent.[54]

On Saturday morning, two faculty of the department of geophysics at the Colorado School of Mines opened the Mansfield Oil Finder and GeBauer's own wonder machine. They were prepared to spend the weekend deciphering complicated circuitry, but there was none. The professors told the jury that the Mansfield detector contained only coils of wire connected to a voltmeter. GeBauer's machine had only a few feet of wire connected to the dials on the face of the box, and nothing else. Professor George T. Merideth testified that the machines that GeBauer submitted were incapable of detecting anything.[55]

The jury took less than three hours to find both men guilty on all counts, which could carry prison terms of up to thirty years, though the probation department recommended probation as the only way Newton and GeBauer could repay Herman Flader. In June 1954, Judge Frank Hickey sentenced Newton and GeBauer to five years of probation, on the condition that they repay $79,452 to Herman Flader, and $2,734 for court costs.[56]

Expecting Silas Newton to repay his victim was a pipe dream. He had testified that he had no income for the past fifteen years, and that all his and his family's expenses were paid by his company. Newton and his wife lived in a premier hotel, drove new Cadillacs (he bought more than one Cadillac per year), dressed and ate well, and joined expensive clubs—all paid for by the Newton Oil Company.[57]

Newton and GeBauer split up angrily a few years after the trial. GeBauer, whom Newton had once defended as a brilliant scientist, he now called a "professional liar" whose only higher degree was a mail-order diploma from a naturopathic school in Texas. GeBauer died in Colorado in 1982.[58]

Newton never paid any of the mandated court costs or restitution. He moved to Phoenix, Arizona, in 1955. Although his home was in an upscale neighborhood of North Phoenix, he insisted that he had zero income and that all his expenses were paid for by his companies. In June 1955 and again in December 1956, Newton's probation officer asked the judge to revoke Newton's probation. A judge recognized in 1957 that Newton was more likely to buzz the courthouse in one of his flying saucers than to pay what he owed. When Newton and GeBauer finished their probation in 1959, GeBauer had paid $4,458. Newton had not paid a cent. The court required them each to sign an agreement to make annual payments of $236, which would have paid off the debt in 329 years.[59]

Together with a couple of other old sharpers, Silas Newton promoted the Tennessee Queen and the Yellow Cat uranium mines in Utah. Investors sued in 1955 after losing $1,235,000. The company prospectus lied, stating that the Tennessee Queen had shipped a hundred tons of high-grade ore and that two hundred drill holes had defined more than nine million dollars of additional ore. A federal grand jury indicted Newton and his partners on thirteen counts of fraud. Newton swore that he was a pauper and asked the court to pay expenses of summoning defense witnesses. The defendants blamed one another; the jury couldn't decide who had done what and found all defendants not guilty.[60]

When Cahn's article, "Flying Saucer Swindlers," appeared in the August 1956 issue of *True*, Scully and Newton sued J. P. Cahn, *True* magazine's publisher Fawcett Publications, editor Ken Purdy, Herman Flader, Denver optometrist Alfred Kleyhauer, and nine officers of Fawcett, for libel.[61] They asked for three million dollars each in actual damages, plus another three million dollars each in punitive damages, for a total of twelve million dollars.

The two men filed suit in Arizona, where Newton lived. All the defendants save Fawcett Publications protested being tried in a state where they didn't live or do business, and which had no connection to the supposed libel. The court agreed, and in March 1958 dropped thirteen defendants from the lawsuit, leaving only Fawcett Publications. Nothing happened until the court dismissed the case in December 1961. Silas Newton later wrote that he dropped the suit because he couldn't pay the legal fees.[62]

Bank president Thomas Bein had given Newton $6,000 for an interest in the Silver Star Mining Company, which owned the Ramsey mine in Arizona, after Newton told Bein that it had stockpiled ore worth more than $1.5 million. Bein's $6,000 was all that he needed to start shipping ore, which would bring Bein a profit of $75,000 in the first year with a total profit exceeding $750,000.

Of course, an unexpected circumstance delayed ore shipments and required more money. There was always one more hurdle that required more money. Bein sued Newton in 1959, but not before he had given him another $19,000. The mine never shipped ore. The case dragged on until Newton settled in 1962 by promising to repay Bein's $25,000 sometime in the future.[63]

Newton's wife Sharon divorced him in 1961. Newton moved back to Denver then to Los Angeles. He kept a home in Sedona, Arizona, close to his Arizona oil swindle.

Newton's Imaginary Arizona Oil Field

Newton went wildcatting near Sedona, Arizona. The nearest oil fields were 200 miles away, but his secret oil-detecting doodlebug told him that there was an oil field. Did any of his friends care to join him in drilling a well and getting rich? It was huge, another Rangely.

Despite Newton's highly publicized dishonesty, he could still coax money from those who should have known better. A perceptive journalist described him as having "the charm and some of the past performance of an O. Henry rogue." Together with former US congressman Richard Harless, Newton formed the Cottonwood Oil Company, and drilled a test in 1959. It was a dry hole, but they never plugged the well, and told people privately that they had discovered oil.[64] Newton and Harless spent all of Cottonwood Oil's capital, so they created the Yavapai Oil Company to drill more wells. In 1961 they drilled the Yavapai Oil #9 Federal, nine miles north of the first dry hole. Newton again told investors that he had found oil but kept it quiet so as not to alert competitors.

Newton's money sources were becoming as dry as the wells he drilled. He needed oil—not much, just a bottle full, as seeing is believing. In 1963, Newton and Harless gave a Phoenix couple a bottle of crude oil from one of the wells. The couple paid $16,000 for an interest in a lease, then turned over another $3,000. Three years later, they sued Newton and Harless for fraud. A laboratory reported that the sample was not crude oil, but a distilled product similar to diesel fuel.[65]

A Yavapai Oil employee gave a sample of supposed crude oil from the #9 Federal to an employee of the Arizona Oil and Gas Commission. The commission sent it to a laboratory, which found that it was a refined product, not crude oil.[66] The commission demanded an oil production test of the well under their observation, but Harless stalled.

Creditors foreclosed on Newton's home in Sedona. An unpaid oil field contractor filed a lien for $1,500 against Newton. Harless and Newton broke up their partnership in 1965, and Harless kept control of the oil properties. Newton told psychic investigator Harold Sherman that he had found a great oil field in Arizona but was keeping it secret while he raised money to gain control of the whole thing. Harold Sherman jumped at the opportunity to invest.

After years of excuses, Newton's investors urged him to produce oil from the wells already drilled. Newton angrily rejected this as thinking small. Producing his wells would alert competitors. If his investors would be patient and trust a little more of their savings to Silas Newton, they could all get rich.

This was Silas Newton's genius: stringing suckers along and getting them to throw good money after bad. The correspondence shows Sherman and friends standing loyally by Newton while he swindled them amid profound expressions of friendship. He always needed a little more money, a little more patience, to overcome one final obstacle. They did not want to admit to themselves that they had been suckered and they kept giving him money.[67]

Newton told investors that he would develop his secret oil field with profits from his fabulous Buckhorn mercury mine near Silver City, New Mexico. He just needed a little startup capital to get the mine running. Newton passed around an assay that showed that the rock was 20 percent mercury—very rich ore. Newton's investors were highly impressed but did not realize that assayers only analyze what is handed to them; they do not know where the sample was taken, or whether or not the sample was salted.[68]

Newton relayed to his investors that once he was ready to ship mercury ore from his mine, he discovered that no one wanted to buy his ore, and the only solution was to own a concentration mill. His investors supplied money to buy an old ore mill north of Deming, New Mexico. Newton supposedly arranged to treat the ore with a brilliant new process, but a year later, the mill was still idle, forcing Newton to return to the old but proven flotation process. When his mill was ready to produce mercury concentrate, Newton told investors that no one wanted to buy his mercury concentrate, and that he would have to build his own smelter.

The smelter was ready by September 1968. A man in the office adjoining Newton's in Silver City later told the FBI that he had overheard Newton planning a trip to El Paso to buy some mercury—enough to show investors as the recovery of a day's test run of his mill and smelter. Around this same time, Newton announced that his mercury mine was also one of the world's great silver deposits, with $10,000 in silver to the ton, promising investors that in three months the profit from the silver mine would give them enough money to save the Arizona oil wells.[69] This meant delay while the mill and smelter were redesigned for silver.

Investors in Newton's Arizona oil wells complained to the Los Angeles district attorney, who charged Newton with grand theft and selling securities without a license. When Newton showed up for a hearing in December 1967, the judge ordered him put in jail. Newton fell ill and landed in the jail hospital. His friends, most of whom were also his victims, and his ex-wife Sharon rushed about to find $2,500 bail and a guarantee of $12,500 before the authorities would free him. He left jail on December 22 after two weeks. The district attorney apparently didn't want an ill eighty-year-old man in jail and dropped the charges after Newton repaid the plaintiff.[70]

Newton neglected to pay the small rental for the Arizona lease beneath his supposed oil discovery wells, the #9 and #27, and the federal government terminated the lease. Newton was running out of excuses. He would finance his oil wells with revenue from his mine but needed money for mine development;

then he needed an ore mill; then he needed a smelter. Now he decided that he needed his own source of natural gas to fuel the smelter. Fortunately, he knew just where to drill, because his oil-finding machine detected gas near the ore mill, a couple of hundred miles from the nearest oil or gas well.

The well started drilling in January 1972, hit basement rock at 7,750 feet, but kept drilling to 10,500 feet, more than two thousand feet into the Precambrian basement. The well reported "oil-saturated sand," as did so many of Newton's dry holes. He set casing and perforated from 9,024 feet to 10,002 feet in Precambrian granite, but the formation yielded only water.

Silas Newton died in Los Angeles on December 15, 1972, age eighty-five, and everything fell apart. Banks called in loans, creditors filed liens, and investors sued. Newton's investors in the New Mexico well believed that because of the "oil-saturated sand" they had an oil well. After Newton died, they tried to produce oil, but just wasted more money. A second well, a mile from Newton's, was drilled on the basis of the supposed oil-saturated sand. It was, of course, another dry hole.

Newton's ex-wife Sharon came forward with his handwritten will making her the executor and principal heir, but inheriting Newton's estate was an honor without profit. Newton owned nothing of value. There was no oil in his Arizona and New Mexico wells. There was no mercury and no silver in his Buckhorn mine.[71] His assets totaled $16,000. There were 140 claims totaling $1.35 million filed against his estate.[72]

In 1998, a person approached UFO researcher Karl Pflock with twenty-seven handwritten pages by Silas Newton himself. The source insisted on anonymity and allowed Pflock to study the document and take notes in his presence. Pflock compared the handwriting with Newton's handwritten will, and they seemed to match. Pflock studied the document on three occasions before the source broke off contact.[73] In the manuscript, Newton bragged that the Aztec crash was a scam masterpiece. Pflock thought that the memoir was probably genuine, but he hesitated to trust the word of a self-admitted con man. Was this the truth, or was it Silas Newton's last con?

Silas Newton never got far with his doodlebug from outer space, but his imaginary saucer landing near Aztec succeeded tremendously. The crash continues to generate books, videos, and podcasts. Newton and GeBauer continue to con people into believing their bad science fiction.

THE SNIFFING PLANES: SWINDLING A LARGE OIL COMPANY

Doodlebugs are not the sole province of the unsophisticated. With the right scientific patter, and especially with the right connections, doodlebuggers can dupe executives and scientists of major oil companies. One such pair took the French oil company Elf and the French government for hundreds of millions of dollars. They were the eccentric Belgian Count Alain de Villegas and Aldo Bonassoli, an Italian television repairman.

Count Alain de Villegas de Saint-Pierre Jette was born to Belgian nobility and grew up in the family château north of Brussels. His mother was a spirit medium who communicated with her ancestors. De Villegas believed that the world would end in the year 2000, but that space aliens would save selected people. He gained an engineering degree, installed a laboratory in his mansion, and labored to transform lead into gold; not to swell his fortune, but to lift third world countries out of poverty.[74]

Aldo Bonassoli was a television repairman and telephone-company electrician but advertised himself as a "professor of nuclear physics." He was born and grew up in Lurano, northern Italy. In the early 1960s, he invented an "electronic telescope" to take detailed pictures of satellites passing overhead, but the satellite images he showed reporters were copied from science fiction comic books. Bonassoli also invented a "death ray" of previously unknown omega subatomic particles that disintegrated solid objects. Bonassoli demonstrated his death ray successfully to reporters by disintegrating a rock at a distance of sixty feet, and a mannequin, aluminum boxes, and a steel bar at seventy-five feet. However, in later demonstrations, perhaps due to increased scrutiny, the death ray failed.[75]

Bonassoli met de Villegas in 1964, introduced by Marco Tedeschini who founded the International Academy of Psychobiophysics. Tedeschini was an engineer turned scientific crank who had devised psychobiophysics as a "unified theory of everything," the sort of mystic science that appealed to de Villegas. Tedeschini introduced his pupil Bonassoli to the count, and the odd couple hit it off immediately. De Villegas thought that Bonassoli's electronic telescope was the work of a genius. The mayor of Lurano warned de Villegas that Bonassoli was untrustworthy, but de Villegas replied that Italians did not appreciate genius.

De Villegas hired Bonassoli as the ideal partner in invention: de Villegas would dream grand ideas and Bonassoli would figure out how to build them.

De Villegas wanted to make fresh water from sea water by bombarding the water with particles to precipitate the salt. After Bonassoli spent months working out the details in the laboratory of de Villegas's Belgian castle, they moved to field testing on the Spanish island of Ibiza. The experiment ended badly when investors caught Bonassoli using tap water to fill the tank of supposedly desalinized sea water.

De Villegas's wealth and title gave him entrée into a network of influential and conservative European Catholics, including Jean Violet, lawyer and specialist in developing inventions. De Villegas was working on a machine to detect underground water, and Violet agreed to promote the invention in exchange for a share of the profits. In 1969, Violet introduced de Villegas to Carlo Pesenti, an Italian financier who needed water for his cement plants, and agreed to fund de Villegas to perfect the water-finding machine. De Villegas still regarded Bonassoli as a technical wizard and hired him to work on the aquifer-detection apparatus. Although the Italian invented, built, and operated the various machines, the Belgian count represented himself as the inventor and Bonassoli as only a technician. The search for water to supply Pesenti's cement plants resulted in only dry holes. Bonassoli made excuses, but Pesenti had seen enough.

The rapidly rising price of crude oil in the 1970s was shocking the world economy, so Bonassoli announced that the water-finding machine could also detect oil. Pesenti was skeptical but agreed to a test. De Villegas put his apparatus on a truck and drove around Holland with two of Pesenti's engineers. The machine beeped every time the truck passed a gasoline station, and the machine even gave the octane rating of the fuel. De Villegas and Pesenti flew the machine over North Sea oil fields. Pesenti agreed to more funding but told de Villegas that the real test would be to find a new oil field. De Villegas feared that if he found oil in areas dominated by big oil companies they would steal his invention, so he decided to search in South Africa, a nation without known oil deposits.

Bonassoli developed two instruments using his new particle and gravity waves. The "Delta" could detect underground oil even from an airplane at an altitude of 20,000 feet. The second invention, the "Omega," was taken to the site in a truck or van, and images on a video screen would tell them the depth, thickness, and lateral extent of the aquifer or petroleum reservoir.[76] Bonassoli called his Omega apparatus *"l'oeil"* (the eye).

De Villegas's daughter and a Belgian pilot smuggled the Omega into South Africa and drove around looking for oil. They found some in Zululand but needed South African permission to overfly the area to define its limits. Violet's friend Antoine Pinay, a former French finance minister, convinced the South African government to approve the flights.

South Africa was desperately short of oil and welcomed the chance to see the revolutionary technology. Pesenti had Bonassoli's Delta mounted in a DC-3, which crisscrossed Zululand, and found the best place to drill. The hole drilled deeper and deeper without finding oil and was plugged at the end of 1975. After it was plugged, a new survey by Bonassoli showed that the borehole had just missed the oil. Pesenti was unimpressed and cut off funding.

The South Africans suggested that they test the system over the abandoned Ogies coal mine, which South Africa filled with crude oil as its strategic reserve. The machine detected oil with an accuracy of 71 percent on an east-west overflight, and 31 percent flying north-south. These results were no better than chance, and the South Africans concluded that the system didn't work. But de Villegas and Violet reasoned that a machine that could detect oil with 71 percent accuracy was a great improvement over existing methods. By this rank statistical fallacy, they concluded that the Delta was a brilliant success.

Elf and the Will to Believe

Violet was a friend of Jean Tropel, the head of security for the French national oil company, Elf. Violet also called on two influential associates: President of the Union Bank of Switzerland Philippe de Weck, and former French Finance Minister Antoine Pinay. They introduced de Villegas to the leadership of Elf.[77] Count de Villegas feared that a big company such as Elf would steal his invention, and he insisted that the number at Elf who knew about the invention be very restricted. He especially didn't want scientists near his machines.

France was suffering from the success of the OPEC cartel in raising the price of oil, and the government pressed Elf to increase production, but Elf had just lost its oil properties in Iraq and Algeria, and had less oil, not more. So when Pinay, Violet, and de Weck told Elf CEO Pierre Guillaumat that their friend Count de Villegas had a breakthrough oil-finding device, they found Elf with a need to believe.

Elf loaded the Delta device into a DC-3 and flew over four small oil and gas fields in the Aquitaine region of southwest France. De Villegas was at the front of the plane, and Bonassoli behind with the Delta. The weather was clear, but windows were shut to prevent Bonassoli from recognizing landmarks. Each time the plane flew over one of the oil and gas fields, the Delta beeped obediently, and Bonassoli gave a correct estimate of the depth to hydrocarbons at each one. The plane flew over a geologic structure known to be barren: the Delta remained silent. On the return, they overflew a nearly exhausted gas field, and the Delta responded feebly.

The Delta performed impressively. But de Villegas and Bonassoli had their other revolutionary device: the Omega. The Omega had shorter range but was more accurate. Elf mounted the Omega machine on a pickup truck, and with Villegas and some Elf geologists in the cab and Bonassoli in back operating the machine, they rode around while the machine accurately delineated the depth, lateral limits, thickness, and oil content of the Lannemezan and Bonrepos oil fields.

At Castera Lou the Omega astounded Elf geologists. The oil field had been discovered less than a month before, and the details were known only to Elf insiders. Bonassoli not only told them the depth and thickness of the oil deposit, but zoomed in on the discovery well, and showed the borehole branching where the driller had pulled back and sidetracked. And there on the Omega screen appeared a bit of junk in the abandoned branch of the hole: a metal bar with a nut. The Elf people were astonished. No one outside of Elf could have known such a detail.

Journalist Pierre Péan wrote that following the Aquitaine Basin test of 1976, Elf employees were like "zombies," their critical sense blinded by the incredible possibilities of the Delta and Omega. Here were oil-finding instruments far superior to any other, a giant leap ahead of anything they or their competitors had, or even had imagined. With such miracle oil finders, Elf could leap over its much bigger American and British rivals and dominate the oil industry. The nation could not only end its oil dependence on unreliable countries, but France would become an oil powerhouse and assume a greater role in world affairs. The possibilities were intoxicating. An Elf employee later recalled: "Anybody who criticized was practically suspected of being a bad Frenchman."[78]

During the next three years, when Bonassoli's oil-finding gizmos failed repeatedly, Elf geologists would remember the amazing performance at Castera Lou, and believe that there was something to the mysterious technology. Péan concluded that Bonassoli must have obtained inside information on Castera Lou from an Elf employee.[79]

It's Who You Know

Had Aldo Bonassoli walked into Elf headquarters to sell his oil-finding wonders, he probably wouldn't have got past the receptionist. And even if he convinced some low-level managers, Elf would have looked into Bonassoli's background before paying him. That his oil-finding scheme was taken so seriously, and lasted so long, was due to concentric spheres of power and influence surrounding the Delta and Omega. Despite Bonassoli's past deceptions, Count de Villegas vouched for him and, more importantly, de Villegas presented himself as the inventor and Bonassoli as only a technician. De Villegas, in turn, was regarded as eccentric but honest, a good conservative Catholic, and so gained the support of Jean Violet. Finally, Violet was a person of undoubted judgement and integrity who could enlist such luminaries as former Finance Minister Antoine Pinay and Philippe de Weck, president of the Union Bank of Switzerland (UBS). He was also a friend of the head of security at Elf, Jean Tropel.

With an introduction by Tropel and their own prestige, Violet, Pinay, and de Weck could go directly to the president of Elf, bypassing preliminary tests by field geophysicists, and making Elf's evaluation depend on a single field test, which was somehow rigged. And with such prestigious people backing the machines, Elf thought that they had no reason to investigate the principals behind the Delta and Omega.

De Villegas created a company named FISALMA, which consisted of Panamanian incorporation papers and an account in the UBS. In May 1976, the chairman of Elf went to UBS headquarters in Zurich and signed a contract that gave Elf exclusive use of Delta and Omega for one year, in exchange for four hundred million French francs (eighty-five million US dollars).[80]

Eighty-five million dollars! Most doodlebuggers are lucky to snag a few thousand here or there. De Villegas and Bonassoli had just left every previous doodlebugger far behind in the dust. We can see why they dealt with a large oil company like Elf. It's the same reason that Willie Sutton robbed banks: that's where the money is. And that was just to start, just for one year's use of the unproven device.

Inspired by the boundless potential of Delta and Omega, de Villegas spent his Elf money on two creations: the Centre de Recherches Fondamentales (CRF) and the Compagnie Européenne de Recherches (CER). He renovated the family estate of Rivieren in Belgium, where he installed CRF, which had twenty employees with laboratories and computers. CER built a 2,000-square-meter airplane hangar at a military airport and filled it with four airplanes, including a Boeing 707, twelve pilots, and thirty ground workers. The company also had a modern twenty-meter boat for prospecting the ocean. De Villegas also spent $2.8 million building a church in the south of France and donated seven million dollars to Catholic aid projects in Africa and Spain.

Elf had Bonassoli look for oil in some areas Elf scientists considered prospective. They flew Bonassoli and his Delta machine in clandestine flights over France, Spain, Portugal, Switzerland, Netherlands, Ireland, and Brazil.[81]

The one-year contract was over before Elf could prove the miracle machines with new oil discoveries. The Montegut 1 started drilling in February 1977 and was still drilling in the Aquitaine Basin with no sign of oil when the contract expired. A new Zululand well in South Africa began drilling in May 1977. Elf renewed the contract with de Villegas and Bonassoli in June 1977: another four hundred million francs (another eighty-one million dollars).

The Montegut stopped drilling in September 1977, a 4,483-meter dry hole. The Omega now showed that the borehole had just missed the oil reservoir.

De Villegas and Bonassoli refused to allow Elf scientists to look inside their machines. When the scientists pressured them about how the machines worked, Bonassoli told them: "Don't try to understand what you will see."[82] When Elf questioned why the machines failed to find oil, de Villegas threatened to sell them to some big American oil company. Bonassoli would amaze visitors to de Villegas's Belgium laboratory by taking a book or a piece of paper to an adjacent room and press it against the wall opposite the Omega. Bonassoli would focus the Omega machine, and an image of the page would appear on the screen. Elf geologists were so convinced by the May 1976 demonstration that they ignored signs of fraud on later flights.

Despite two years of failure, Elf agreed in June 1978 to give the inventors 250 million Swiss francs ($134 million US dollars). But this time Elf insisted that de Villegas and Bonassoli explain the theory and allow Elf scientists to test the devices. The two agreed, as long as the investigators did not peek inside the equipment.

The Delta and Omega predicted three more places with oil, which produced three more dry holes. An extension to the Castera Lou field failed. A wildcat well in the Mediterranean was also dry, as was another offshore wildcat in the French Atlantic. Elf bought leases to drill still more Bonassoli prospects in Gabon, Morocco, and Spain. Despite the agreement, de Villegas and

Bonassoli refused to explain the apparatus beyond generalities. They said that they were no longer interested in Elf, because they were negotiating with Chase Manhattan, a bank with historic ties to Exxon, for far more money.

All during the negotiations, drama, and dry holes, Elf's Zululand well in South Africa kept drilling deeper. The company finally abandoned the hole in December 1978, in basalt, after drilling 6,083 meters (19,957 feet). It was a hugely expensive failure and Elf demanded that the inventors reimburse the company for the cost. The inventors had promised to do so for any dry holes, but this was the first time Elf asked for dry hole costs. Jean Violet agreed with Elf that the inventors should pay, which de Villegas considered a stab in the back.

No longer trusting Violet, de Villegas brought in an American, Daniel Boyer, as his advisor. Boyer had a resume out of a spy thriller: born in Yugoslavia, he fought in the resistance against the Nazis and was captured, tortured, and imprisoned in an abandoned coal mine. He tunneled his way out and made his way to France, where he adopted the French name Boyer and fought in the resistance, ending the war as a second lieutenant—while still a teenager. After the war he moved to the United States. Boyer had connections to top American politicians, and some suspected that he was a CIA agent. The French feared that he would pass the secrets of Delta and Omega to US oil companies. De Villegas was always suspicious of the French and kept the machines in his Belgian castle. He hired a private security firm but discovered that the guards were allowing Elf employees secret after-hours access to the offices.[83]

Elf turned to Bonassoli. Despite the count's insistence that he himself was the inventor, Elf personnel realized that Bonassoli was the real one. High-level Elf management met with Bonassoli, flattering him as the real inventor and telling him that he didn't need de Villegas. Bonassoli loved being addressed as "professor" by graduates of the highly prestigious École Polytechnique (Polytechnical School). As the crowning gesture, on April 5, 1979, President of France Valéry Giscard d'Estaing came to observe a test of the Omega at the Soudron oil field, after which Giscard spoke warmly to Bonassoli.

Bonassoli enthusiastically agreed to work directly with Elf, telling them how de Villegas insulted and mistreated him. But getting a coherent explanation still proved impossible. Two young Elf geophysicists were assigned to Bonassoli, who told them that he had discovered a new subatomic particle, a type of neutrino, which retained a record of all the materials it passed through. The French called the particles "aldinos," after Aldo Bonassoli. But that was about all they learned. They reported that Bonassoli seemed to have a good grasp of physics, but his explanations were opaque. For one thing, Bonassoli did not speak French well: his speech came out in disconnected phrases, impossible to follow into complex concepts.

After Bonassoli successfully demonstrated the Omega to President Giscard, Elf arranged for the minister of industry to observe the Omega in action at a test at the Lacq gas field. But instead of the expected triumph, it was a disaster. Minister André Giraud recalled: "It was obvious to the naked eye that it was a joke. There was not the slightest doubt in my mind." He refused to stay for the lunch in his honor with the inventors. Giraud ordered a controlled test of the Omega under the supervision of Jules Horowitz, a top scientist at the French Atomic Energy Commission. The test took place May 24, 1979, at an Elf facility outside Paris.[84]

Back in the Belgian laboratory, Bonassoli had demonstrated that the Omega could detect objects and even read from books through a solid wall, so Horowitz put a complex metal object on the side of the wall opposite the machine. The equipment failed, and Bonassoli made up an excuse. A second test, using an object inside a sealed envelope, also failed—another excuse. After discussion, Bonassoli agreed that a metal ruler would provide a fair test, and Horowitz took the ruler to the other room and placed it against the wall. This time a long rectangle appeared on the video screen. But Horowitz returned and showed that before putting the ruler against the wall, he had bent it into a V. The Omega was an obvious sham.[85]

Elf technicians opened the Omega machine and found a photocopy of a long rectangle identical with the image on the screen. It was obvious that the video screen could only show images that had been preloaded into the Omega. The same was true when they examined the airborne Delta machine: no apparatus to receive an outside signal, no computer to process the information, just a video screen set to display paper images preloaded into the machine.

Yet even being caught red-handed was not the end. Bonassoli said that he had failed the test on purpose, and some Elf scientists were so invested in the machines that they still believed in it and pushed for another test. All parties agreed on a time, place, and test protocol on June 27 at Wolfsberg castle. Bonassoli began the test, but after two hours of adjustments, he announced that the Omega was not working, and so the test could not be done. Elf technicians, who had studied how Bonassoli operated the machine, took over, and after two more hours of adjustments, the Omega started working. But a pattern appeared immediately on the screen, before an object could be placed behind the wall. The technicians opened the machine and found a paper image of the same pattern. The test continued but did not succeed. Bonassoli had again been caught cheating. Elf immediately demanded that the assets of FISALMA and other doodlebug-related entities be frozen, and both de Villegas and Bonassoli agreed.

Elf stopped a planned payment of $110 million to Bonassoli and de Villegas and recovered $100 million in escrow at UBS. De Villegas disappeared. Bonassoli scurried back to Italy and denied that he had swindled anyone. He insisted that he had deliberately sabotaged the test because he had not been fully paid. Investigators confiscated the Omega machine. Inside they found a collection of images in various simple shapes.[86]

French investigators judged de Villegas to be "a little crazy" but honest and put all the blame on Bonassoli. Elf clawed back much of the money paid to de Villegas, but the French *Cour des Comptes* (Court of Auditors) estimated the total losses, including dry holes, as between 740 and 790 million francs (about 160 to 170 million dollars).[87] The government declared the useless equipment to be a military secret. The chief government accountant was sworn to secrecy, destroyed his copies of the report, and the only surviving copies were removed by President Giscard and Prime Minister Barre when they left office in 1981.

The scandal seemed successfully buried until tax accountants for the French government noted extraordinary payments by Elf of five hundred million French francs to the Union Bank of Switzerland. In March 1983, the president of Elf's parent company ERAP was stunned by a letter from the finance ministry demanding repayment of the five hundred million francs sent to Switzerland. People began asking questions, and the strange tale emerged.

The Chained Duck

Although the new president, François Mitterrand, had no affection for former President Giscard, the Elf investigation remained within the government until word leaked to the satire magazine *Le Canard enchaîné* (The Chained Duck). *Le Canard* is famous for digging out political scandals, which it then mocks with biting wit. *Le Canard* exposed the scandal in June 1983, calling the petroleum detectors *avions renifleurs*, "sniffing planes," and revealing that a pair of oddball foreigners had fooled the nation's top politicians and scientists, including elite graduates of the École Polytechnique, into wasting hundreds of millions of dollars for a pair of photocopiers. The story caused a frenzy, as other news outlets rushed to catch up and to provide details.

Ex-president Valéry Giscard d'Estaing defended his actions in a dramatic television appearance and accused Mitterrand of betrayal. He argued that each French leader had a duty to rise above politics and cover up past scandals. But of all the perils of politics, ridicule can be the most fatal. It was difficult to justify spending large sums for something known as the farcical "sniffing planes," and the affair dimmed Giscard's prospects for a comeback.

Count de Villegas hid from reporters by going to South America to build landing strips for flying saucers. His château was burglarized and his files rifled. Belgian police said it was the work of professionals. Count Alain de Villegas died a ruined man.[88]

Bonassoli told reporters that his oil detector worked, but that he had sabotaged the test because he had not been paid. He was perfecting an improved version, and if the French and Italians spurned his invention, he would take it to the United States. Daniel Boyer and Jean Tropel believed that Bonassoli could not have fooled so many smart people for so long unless he had something. They kept in touch with the Italian while he tinkered, but they finally gave up in 1981. As of 2013, Aldo Bonassoli was earning a modest living as a television repairman.

The affair of the sniffing planes is unique in the great sum of money extracted from its victim, but perhaps it is not as unique as it appears. When a company is swindled and it cannot recover the money, the human instinct is to bury the matter and hope that the public—especially shareholders—does not learn how company management was fooled. The sorry saga of the sniffing planes came very close to being permanently hidden. It makes one wonder, how many other companies have been taken by doodlebug swindles, but succeeded in hiding the fact?

LOOKING BACK, LOOKING AHEAD

NO MATTER WHAT YOUR OPINION ON DOWSING AND DOODLEBUG-ging, there is much to admire in the optimism, ingenuity, and originality shown by unconventional oil finders through the years. In contrast to the meager mention in the oil literature, the United States has a rich history of oil dowsing and doodlebugging that merits greater recognition.

Figures 26 and 27 show the number of documented active dowsers and doodlebuggers in the United States each year, compiled from industry literature and popular news media. Included in the counts are 186 doodlebuggers and 272 oil dowsers.[1]

Compared to contemporary estimates, the numbers of dowsers and doodlebuggers on the chart are at least an order of magnitude too low. Despite limitations of the sample, the charts are assumed to show broad trends spanning the time period. Of particular note is the fact that the trends of doodlebuggers and oil dowsers are very different.

The chart in figure 26 shows the numbers of dowsers. The observation is inescapable that oil dowsing has become increasingly rare in the United States during the past fifty years. This is the subjective impression of people in the industry, supported by the dwindling number of oil dowsers mentioned in industry literature and popular media. Oil dowsers are progressively older and fewer.

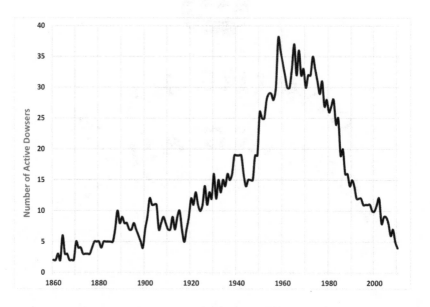

Fig. 26. The number of known active oil dowsers in the United States based on industry literature and media reports.

The rise of geophysics severely reduced the number of pseudo-geophysical doodlebugs from their peak in the 1920s, but doodlebugs live or die on their scientific pretensions—which geophysicists are able to refute. Dowsers, on the contrary, are seldom glued to a theory. They have some ideas, but pressed for proof, they are likely to shrug and say that it just works, and the proof is in the drilling.

The decline of oil dowsing might be due in part to reduced numbers of independent oil operators, and by changes in tax law. The small operator has always been the dowser's most likely client. From the 1940s through the 1970s, the much higher marginal tax rates for ordinary income, compared to the rates for capital gains, created a powerful incentive for high-salaried people to invest in oil and gas drilling. Many small independent oil operators funded their drilling with tax-shelter-seeking investors. The reduction in peak marginal tax rates from 91 percent in 1964 to 33 percent in 1988 eliminated much of the advantage in oil-drilling tax shelters and deprived many oil dowsers of their clients.

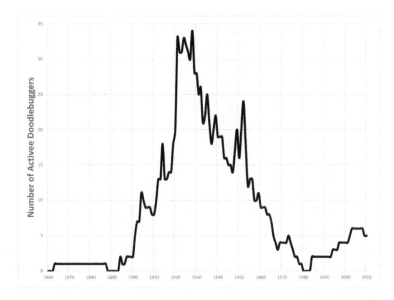

Fig. 27. The number of known active oil doodlebuggers in the United States based on industry literature and media reports.

The Future of Dowsing

"Passion and mockery is still the current fate of dowsing."

—Anne Jaeger-Nosal[2]

Passion is on the side of dowsers, mockery on the side of skeptics. Most (not all) scientists say that not only doesn't it work, it is theoretically impossible for dowsing to work. Dowsers respond that dowsing is simply a fact, and fact trumps theory. Scientists point to controlled tests that show only chance rates of success. Dowsers shrug and say that dowsing succeeds in real-life situations, and if tests are unsuccessful, that is the fault of the tests. Scientists and dowsers rarely try to convince the other side of their error.

There do not appear to be any studies under controlled conditions that unequivocally prove that dowsing works. In the few studies for which a dowsing effect is claimed, the improvement over chance is so small that the argument descends into arcane questions of statistics. Dowsers generally argue that

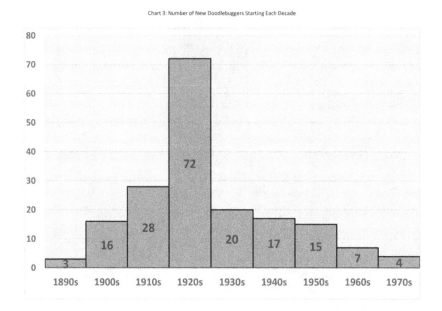

Chart 3: Number of New Doodlebuggers Starting Each Decade

Fig. 28. The number of new oil doodlebuggers becoming active in each decade in the United States.

controlled laboratory experiments cannot replicate field conditions. But using field results introduces many more variables, making it much more difficult to prove or disprove dowsing. Anecdotes that this or that well or handful of dowsed oil wells were successful are statistically insignificant. The available data indicate that, for most dowsers, even some highly celebrated ones, the results are indistinguishable from chance.

The search for paranormal phenomena is sometimes likened to a search for a white crow. No matter how many black crows you find, you can never prove that all crows are black. In the absence of a controlled test of oil dowsing, we are left with statistical comparisons of success rates of dowsing versus geology.

Doodlebugs in the Era of Geophysics

Figures 27 and 28 show, respectively, the number of active doodlebuggers in the United States each year and the number of new doodlebuggers starting their careers in each decade. The 1920s was the dominant decade for American doodlebugs, with greater than a third of the known doodlebugs starting in that single decade. The 1920s was also the decade when genuine geophysical methods (gravity, seismic refraction, and seismic reflection) were introduced. The great success of geophysics created a cadre of professional geophysicists able to identify pseudo-geophysics, and with enough influence within oil companies that their negative opinion usually keeps a company from using worthless methods.

The era of the self-taught doodlebug inventor is gone. In response to scrutiny from geophysicists, doodlebuggers have adopted the protective coloration of real geophysics, often shielding themselves with an impenetrable fog of geophysical jargon. Not that they are all frauds; indeed, conscious frauds are the exception. Most doodlebuggers today, as in the past, are convinced that their instruments detect oil.

It is sometimes difficult to tell geophysics from pseudo-geophysics. Even oil-industry geophysicists may be fooled, especially those who deal only with seismic geophysics and who may be too specialized to judge other methods. The Technology Assessment Group has prepared a brochure which lists common characteristics that help distinguish good geophysics from bad.[3]

Do not mourn the extinction of the doodlebugs. They are still with us.

BLACK BOXES AND MARTIAN CANALS

"The first principle is that you must not fool yourself—and you are the easiest person to fool."

—Richard Feynman, Nobel Prize winner in Physics[4]

Richard Feynman observed, as had Irving Langmuir (Nobel Prize winner in Chemistry) twenty years before him, that even the smartest scientists can fool themselves. How is this possible?

Scientists Gone Bad: Martian Canals

In 1877, the Italian astronomer Giovanni Schiaparelli looked through his telescope and saw straight dark lines crossing the planet Mars. He called them *canali*, apparently meaning channels or grooves, but the world latched on to the alternate translation: "canals," implying an artificial origin. Soon astronomers worldwide were reporting canals and mapping them in detail.

One of the most prominent canal observers was Percival Lowell, a Bostonian of impeccable social and scientific credentials, who peered at Mars from his observatory at Flagstaff, Arizona. Schiaparelli mapped 113 canals, and the indefatigable Lowell mapped some 300 more. Skeptical astronomers could not see any canals, but there were explanations: perhaps their telescopes were not as good, or the atmospheric conditions less than optimal, or their eyesight not as acute. Mars came into optimal viewing only for a few weeks every few years, and even then astronomers had to wait patiently for air currents to settle, so that—if lucky—they might see clear details of Mars for a few seconds before the currents returned, and the face of the planet returned to a blur.[5] The question was only resolved in 1965 when the American space program received its first close-up photographs of Mars from Mariner IV, launched the year before.

Before Mariner IV smashed into Mars, the photographs it radioed back dismantled the idea of Martian canals. There were no canals. After eighty years of canal mapping, the canals were shown in an instant to be just the junk of imagination, and all the painstakingly detailed maps of Martian canals were thrown in the trash. Schiaparelli and Lowell did not live to see their life works made mockery by the Mars probe. They were scientists of undoubted integrity—who saw what they wanted to see.[6]

Back to the Black Box in Calgary

We return to the black-box doodlebug introduced at the start of the book.[7] Wesley C. Miller, the chief sound engineer for MGM Studios in Los Angeles, designed a machine to find oil; "invented" would be too strong a term, because it was a basic radionics instrument adapted for oil exploration. Miller demonstrated the machine to a geologist for Cities Service and contacted the head of the US Geological Survey but could not interest them. He did interest Calgary oil promoter George Cloakey, who hired Miller to use his gizmo to evaluate drilling locations starting in 1954, but Miller's machine led to a series of dry holes.

By 1960, Cloakey had brought in geologist George S. Hume to evaluate the Miller machine. Hume was former chief of the Geological Survey of Canada and had been awarded the Order of the British Empire for his achievements in developing Canadian oil resources. He authored more than seventy technical papers, many of them award-winning. The *Bulletin of Canadian Petroleum Geology* later eulogized him as "Canada's foremost authority on the petroleum and natural gas resources of our country."[8] He had retired from the Geological Survey in 1956 and moved to Calgary.

Hume reviewed the doodlebug surveys and concluded that Miller's results were too unreliable. Miller's career with MGM (in the 1950s he received four Oscar nominations) left him with little time to search for oil in Canada, so he taught George Hume how to operate the machine. Miller committed suicide in 1962, leaving Cloakey in possession of four Miller machines, each slightly different.

Wesley Miller's oil-finding machine was derived from the quack medical radionics machines built by Albert Abrams in the teens and early 1920s. In Abrams's day, the circuitry was connected to a glass rod moving up and down the patient's abdomen until an increase in friction was felt. Abrams trained some 3,500 doctors in his system, and no doubt many were impressed that they felt the change in friction when the dials were set at the positions appropriate for the symptoms of the patients. Few suspected that the friction change was imaginary. Just as a dowser subconsciously causes the rod to dip, and Lowell imagined that he saw Martian canals, the power of suggestion will cause a radionics machine operator to feel an imaginary change in friction.

In Miller's version, the friction rod was changed to plastic, was mounted on the machine, and instead of doctors fooling themselves, it was a geologist. Hume learned to adjust the knobs on the face of the machine with one hand, while his other hand, wearing a white silk glove, stroked a hard plastic rod attached to the face. When he felt a slight increase in friction in stroking the rod, that increase indicated a change in the rock properties of a type and at a depth determined by the positions dialed on the knobs. He convinced himself that it worked. Because he was highly knowledgeable in geology and petroleum, George Hume made better predictions with the black box than Miller and became convinced that the black box detected the depth and thicknesses of the various geologic formations—and whether they held oil. George Hume had succeeded in fooling himself. Hume was driven around Alberta in the front passenger seat with the instrument on his lap, moving the dials, stroking the plastic rod, and occasionally feeling the change in friction that indicated oil.[9]

George Hume told Dome Petroleum that he had a direct hydrocarbon detector and could prove it. Dome drove him around the East Calgary oil field and noted each time Hume said that they crossed into or out of the limits of the oil field. He was exactly right each time. They stopped at an oil well, and Hume told them the depth of the oil. He was off by only ten feet.

Dome Petroleum flew Hume and the black box to a prospect they were drilling at the Red Earth oil field. Hume set up the black box at the well site, carefully went through the procedures, and told Dome that they would find oil in the granite wash at a certain depth. In addition, as they flew over the east end of Lesser Slave Lake on their way to and from the well site, Hume told them he got a very strong signal for oil and was sure he was over a major oil field.

George Hume was wrong about the well at Red Earth. The granite wash was much deeper than his black box predicted, and the well was a dry hole. Dome geologists also investigated the area around the east end of Lesser Slave Lake but could see no reason to suspect an oil field there. Dome lost interest in Hume's oil-finding machine.

In 1965, Hume pitched his machine to Ned Gilbert, exploration manager for Sun Oil. Gilbert drove Hume out to where Sun had drilled an oil well but had been unable to find an extension of the field. Hume traced the oil reservoir and showed where they could drill more oil wells, but Gilbert could not convince Sun Oil to take the black box seriously. George Hume died a few days later.[10]

After George Cloakey died in 1979, his widow struck a deal with Ned Gilbert that allowed Gilbert to go through the trash bags full of Cloakey's oil business effects, ready to be thrown out. Gilbert salvaged the four black boxes and a stack of George Hume's oil prospects based on the instruments. Gilbert now had a black box, but only a vague idea of how to operate it. He had watched Hume operate the machine, but he couldn't make it work for himself. He showed it to three electronics experts, hoping that they could tell him how to work the instrument, but the experts all told him that the machine couldn't work. Gilbert could find no oil companies willing to drill George Hume's black-box prospects. Ned Gilbert died in 2018.

Langmuir and later Feynman warned how easily even brilliant scientists can fool themselves. Langmuir called it "pathological science." We could say that George Hume fell victim to this, except that in 1964, Chevron—not knowing of Hume's work—drilled at the east end of Lesser Slave Lake and discovered the Mitsue oil field, hundreds of millions of barrels of oil—where George S. Hume and his black box said it would be.[11]

NOTES

CHAPTER 1

1. William Wright, *The Oil Regions of Pennsylvania* (New York: Harper, 1865), 45.
2. Guy Clifton Bell, *Kentucky Petroleum* (Owensboro, KY: Bell Publishing, 1930), 44; *Oil & Gas Journal*, March 8, 1934, 40.
3. Ruth Sheldon Knowles, *The Greatest Gamblers* (Norman: University of Oklahoma Press, 1959), 107–8.

CHAPTER 2

1. "Story of the Coquette Well," *Weekly Observer* (Marcellus, NY), July 8, 1880, 1; J. H. A. Bone, *Petroleum and Petroleum Wells* (Philadelphia: Lippincott, 1865), 72–73.
2. A. C. Kepler, *A. C. Kepler Diary*, Collection of Lancaster County Historical Society, Lancaster, Pennsylvania, 51.
3. Bill Davidson, "Whoosh! Dorcie Calhoun Strikes It Rich," *Reader's Digest* 58, no. 348 (April 1951), 55–58.
4. Charles R. Fettke, "Developments in Pennsylvania in 1950," *Bulletin of the American Association of Petroleum Geologists* 35, no. 6 (June 1951): 1177–87; "Dorcie, the Dreamer," *Newsweek* 37, no. 1 (November 1, 1951), 45.
5. Charles R. Fettke, "Pennsylvania Oil and Gas Development 1951," *Oil and Gas Development in United States and Canada, 1951*, vol. 22, National Oil Scouts & Landmen's Association (1952), 573–82.
6. *Sun-Gazette* (Williamsport, PA), June 5, 1952, 14.
7. *Express* (Lock Haven, PA), September 8, 1975, 3.
8. *State Journal* (Reno, NV), February 28, 1892, 2.

9.   *Forest Republican* (Tionesta, PA), April 24, 1907, 3; *McKean County Miner* (Smethport, PA), February 8, 1906, 1.

10. *Oil & Gas Journal* 18, no. 8 (July 25, 1919): 39–40.

11. *Herald* (Lethbridge, Alberta), December 13, 1919, 12.

CHAPTER 3

1.   Herbert Asbury, *The Golden Flood* (New York: Knopf, 1942), 61–64; *The Derrick's Hand-Book of Petroleum* (Oil City, PA: Derrick Publishing, 1898), 1037.

2.   *Monongahela Valley Republican* (Monongahela, PA), February 28, 1878, 4.

3.   *Weekly Blade* (Keene, PA), June 2, 1881, 3; *Courier* (Waterloo, IA), March 6, 1878, 2; *Herald* (Titusville, PA), August 25, 1879, 3.

4.   Frances E. Willard and Mary A. Livermore, eds., *A Woman of the Century* (Buffalo, NY: Charles Wells Moulton, 1893), 753; Sarah M. Severance, "Elizabeth Lowe Watson," *The Progressive Woman* 5, no. 52 (September 1911), 9; John William Leonard, ed., *Woman's Who's Who of America* (New York: American Commonwealth, 1914), 859; Ernest C. Miller, *Early Daze in Oil* (Philadelphia: Dorrance, 1974), 58.

5.   *Dispatch* (Pittsburgh, PA), May 18, 1891, 5; *The Derrick's Hand-Book of Petroleum* (Oil City, PA: Derrick Publishing, 1898), 380, 495–96.

6.   J. M. Peebles, *The Practical of Spiritualism* (Chicago: Morton & Leonard, 1868).

7.   Chicago *Daily Tribune*, March 13, 1873, 2.

8.   *Herald* (Titusville, PA), July 11, 1866, 3.

9.   *Republican* (Cazenovia, NY), June 12, 1967, 2.

10. John F. Carll, *Oil Well Records and Levels* (Harrisburg, PA: Second Geological Survey of Pennsylvania, 1877), 18–19.

11. *Territorial Enterprise* (Virginia City, NV), March 14, 1868, 2.

12. *Weekly Gazette* (Pittsburgh, PA), November 18, 1868, 7.

13. *New York Times*, November 29, 1884, 3; Harry Botsford, *The Valley of Oil* (New York: Hastings, 1946), 248; Rochelle Raineri Zuck, "The Wizard of Oil: Abraham James, the Harmonial Wells, and the Psychometric History of the Oil Industry," *Journal of American Studies* 48, no. 2 (2012): 313–36. Ms. Zuck leapt to the bizarre conclusion that the oil industry today uses "the combined efforts of science

and psychometry." Psychometry is divining facts of an object by handling it.

14.   "A 'Vitapathic' College," *Medical World* 7, no. 3 (March 1889): 94.

15.   *The Derrick's Hand-Book of Petroleum* (Oil City, PA: Derrick Publishing, 1898), 870.

16.   *Daily Iowa Capital* (Des Moines, IA), October 29, 1897, 9.

17.   *Examiner* (San Francisco, CA), February 20, 1874, 1.

18.   Dan DeQuille, *The Big Bonanza* (1876; rpt. Las Vegas, NV: Nevada Publications, 1947), 62–63.

19.   *Breeze* (Bolivar, NY), November 27, 1896, 1.

20.   *Weekly Tribune* (Hornellsville, NY), December 1,1899, 4.

21.   *Evening Herald* (Olean, NY), May 1, 1920, 3; *McKean Democrat* (Smethport, PA), January 1, 1920, 3.

22.   *Standard Union* (Brooklyn, NY), January 30, 1921, 26.

23.   Los Angeles *Times*, October 25, 1923, part II, 2; June 25, 1924, part II, 22.

24.   John F. Carll, "Preliminary Report on Oil and Gas," *Annual Report of the Geological Survey of Pennsylvania for 1885* (1886), 44.

25.   Barbara Goldsmith, *Other Powers* (New York: Knopf, 1998), 425–26.

CHAPTER 4

1.   *Republic* (St. Louis, MO), January 11, 1903, 52.

2.   *Tribune* (Chicago, IL), October 30, 1902, 2; *Post-Dispatch* (St. Louis, MO), November 9, 1902, 5; *Mid-Week News-Times* (Goshen, IN), August 26, 1902, 3.

3.   *Journal* (Logansport, IN), November 7, 1902, 2.

4.   *Events* (Enid, OK), May 11, 1922, 17; *Banner* (Ocala, FL), July 27, 1906, 9; *Daily News* (Galveston, TX), May 9, 1909, 2; *Weekly Herald* (La Porte, IN), March 3, 1904, 2.

5.   *Daily Star* (Sandusky, OH), February 11, 1901, 3; *Transcript* (North Adams, MA), February 28, 1901, 1.

6.   Mr. and Mrs. Guy Fenley, interview by W. A. Owens, May 5, 1956, tape 182, transcript, Center for American History, University of Texas at Austin, 3–6.

7.   *Daily Star* (Sandusky, OH) February 11, 1901, 3.

8.   *Daily News* (Galveston, TX), September 11, 1901, 4; February 2,

1902, 4; *Advertiser* (Lafayette, LA), December 28, 1901, 1.

9.   Mody C. Boatwright, *Folklore of the Oil Industry* (Dallas: Southern Methodist University Press, 1963), 17–20.

10.   Dallas *Morning News*, March 4, 1964.

11.   Ibid., February 12, 1969.

12.   Thad Sitton and James H. Conrad, *Freedom Colonies* (Austin: University of Texas Press, 2005), 73.

13.   *Semi-Weekly Light* (Corsicana, TX), October 22, 1929, 7.

14.   *Daily News* (Mexia, TX), February 10, 1924, 7; Gene Fowler, *Mavericks* (Austin: University of Texas Press, 2008), 133.

15.   Doris Hollis Pemberton, *Juneteenth at Comanche Crossing* (Austin: Eakin, 1983), 101–2.

16.   Mary Grace Kettner, undated, unpublished, two-page manuscript, Corsicana Public Library, Genealogy Department, vertical file "Annie Buchanan."

17.   *Daily Sun* (Corsicana, TX), March 3, 1954.

18.   William A. Owens, "Seer of Corsicana," *Southwest Review* (1958): 124–34.

19.   *Semi-Weekly Light* (Corsicana, TX), September 10, 1954, 14.

20.   William A. Owens, "Seer of Corsicana," *Southwest Review* (1958): 124–34.

21.   Jess Stearn, *Edgar Cayce: The Sleeping Prophet* (New York: Doubleday, 1967).

22.   *Earth Changes* (Virginia Beach: ARE, 1963).

23.   Edgar Cayce, *My Life as a Seer* (New York: St. Martin's, 1997), 137; Cayce Reading 442-2, January 17, 1934, text line 25; Reba Ann Karp, *The Edgar Cayce Encyclopedia of Healing* (New York: Grand Central, 1986), 90–91, 100, 103, 484.

24.   Sidney D. Kirkpatrick, *Edgar Cayce: An American Prophet* (New York: Penguin Putnam, 2000), 206.

25.   Cayce Reading 3777-2, report item 1, telegram by Joseph B. Long dated October 12, 1920. Readings and related material available from ARE, Virginia Beach, Virginia.

26.   Cayce Reading 3777-5, Affidavits January 25, 1921, by H. P. Salter and J. D. Vinson.

27.   Sidney D. Kirkpatrick, *Edgar Cayce: An American Prophet* (New York:

Penguin Putnam, 2000), 213.

28.   The Sam Davis #1 well is API # 42-093-80232.

29.   A. Robert Smith, *Hugh Lynn Cayce: About My Father's Business* (Norfolk, VA: Donning, 1988), 46; Harmon Hartzell Bro, *A Seer Out of Season: The Life of Edgar Cayce* (New York: Penguin, 1989), 340.

30.   Edgar Cayce, *My Life as a Seer* (New York: St. Martin's, 1997), 131.

31.   Sidney D. Kirkpatrick, *Edgar Cayce: An American Prophet* (New York: Penguin Putnam, 2000), 226–27.

32.   George R. Froh, "Edgar B. Davis: Wildcatter Extraordinary" (PhD diss., Texas A&M University, 1980), 234; Riley Froh, "The Folklore and Facts Behind the Luling Discovery Well," *East Texas Historical Journal* 17, no. 2 (October 1979): 45–54.

33.   A. Robert Smith, *Hugh Lynn Cayce: About My Father's Business* (Norfolk, VA: Donning, 1988), 46.

34.   Sidney D. Kirkpatrick, *Edgar Cayce: An American Prophet* (New York: Penguin Putnam, 2000), 243.

35.   Ibid., 239. David Kahn placed the incidents of losing tools and dropping granite down the hole at the Sam Davis well at Comyn rather than at San Saba. David E. Kahn and Will Oursler, *My Life with Edgar Cayce* (Garden City, NY: Doubleday, 1970), 71.

36.   Cayce Reading 221-2, Background items 3-5; Sidney D. Kirkpatrick, *Edgar Cayce: An American Prophet* (New York: Penguin Putnam, 2000), 211.

37.   Edgar Cayce Readings 4905-13, January 20, 1925; 4905-20, April 18, 1925; 4905-51, December 29, 1925; 4905-68, January 17, 1927.

38.   *Record* (Clifton, TX), August 28, 1925, 8.

39.   Edgar Cayce Reading 4905-70, April 25, 1927.

40.   Edgar Cayce Reading 195-7, July 7, 1924.

41.   Edgar Cayce Reading 943-5, text line 3.

42.   Edgar Cayce Reading 5628-5, January 30, 1927.

43.   Curtis Wilmott, letter to Edgar Cayce, March 27, 1935, attachment to Edgar Cayce Reading 5628-12.

44.   "Oil and gas lease" between B. F. and A. K. Barnes, and Cecil Ringle, April 19, 1920, Comanche County Recorder, Comanche, Texas, vol. 153, 177–78.

45.   Edgar Cayce Readings 1180-4, October 4, 1924; 1180-5,

October 25, 1924.

46. Edgar Cayce Reading 1180-6, October 21, 1925.

47. Edgar Cayce Reading 2519-6, February 17, 1930.

48. Edgar Cayce Reading 1180-8, June 3, 1936.

49. Bruce A. Black, "Oil and Gas Exploration in the Albuquerque Basin," *Albuquerque Country II*, ed. Jonathan F. Callender, 33rd Annual Field Conference (New Mexico Geological Society, 1982), 313–23.

50. Edgar Cayce Reading 1561-12.

51. Edgar Cayce Readings 1561-13, 1561-14, and 1561-15.

52. Edgar Cayce Reading 1561-19.

53. Edgar Cayce Reading 2091-4, March 20, 1942.

54. Edgar Cayce Reading 3777-1.

55. Edgar Cayce Reading 3777-5, 1920.

56. Edgar Cayce Reading 4906-9, 1924.

57. Edgar Cayce Reading 5628-9, May 4, 1928.

58. Cayce Reading 4203-2, Scout card for United Carbon Co. #1 Bastrop Pulp & Paper Co., 18-21N-6E, Ira Rinehart's Oil Reports, n.d.

59. Edgar Cayce Reading 270-16, April 5, 1928.

60. Edgar Cayce Reading 370-1, July 7, 1932.

61. Edgar Cayce Reading 257-211, September 25, 1939.

62. Kenneth Roberts, *Water Unlimited* (Garden City, NY: Doubleday, 1957), 206–7.

63. "Cayce oil drilling venture - #9028," http://investinenergy.wordpress.com/2009/02/14/cayce-drilling-venture-9028/.

64. Ruth Shelton Knowles, *The Greatest Gamblers* (Norman: University of Oklahoma Press, 1959), 302.

65. Harry Hurt III, *Texas Rich* (New York: Norton, 1981), 53.

66. A. R. and R. B. Buckalew, "The Discovery of Oil in South Arkansas, 1920–1924," *Arkansas Historical Quarterly* 33, no. 3 (1974): 232; Tom Buckley, "Just Plain H. L. Hunt," *Esquire* 67, no. 1 (January 1967): 64.

67. James A. Clark and Michel T. Halbouty, *The Last Boom* (New York: Random House, 1972), 43.

68. Ibid., 92–93.

69. *The History of Hunt Oil Company* (Dallas: Hunt Oil, 1984); Ardis Burst, *The Three Families of H. L. Hunt* (New York: Weidenfeld & Nicolson, 1988), 78; Harry Hurt III, *Texas Rich* (New York: Norton,

1981), 71.

70.   Margaret Hunt Hill, *H. L. and Lyda* (Little Rock, AR: August House, 1994), 35–36.

71.   Ruth Sheldon Knowles, *The Greatest Gamblers* (Norman: University of Oklahoma Press, 1959), 303.

72.   Edwin Shrake, *Land of the Permanent Wave* (Austin: University of Texas Press, 2008), 72.

73.   Ibid.

74.   Jeane Dixon, *My Life and Prophecies* (New York: Morrow, 1969), 95–96.

75.   Denis Brian, *Jeane Dixon: The Witnesses* (Garden City, NY: Doubleday, 1976), 205–6.

76.   Joe E. Guyer, "Joe S. Farmer," *American Association of Petroleum Geologists Bulletin* 82, no. 3 (March 1998): 516; *The History of Hunt Oil Company: Our 50th Year, 1984* (Dallas: Hunt Oil Co., 1984).

77.   Stanley H. Brown, *H. L. Hunt* (Chicago: Playboy, 1976), 203–4; Houston *Chronicle*, November 10, 1985, Business section, 18.

78.   Garland D. Ells and Robert E. Ives, "South Michigan's Oil Bonanza," *World Oil* (October 1959), 136.

79.   *Daily News* (Hillsdale, MI), June 25, 1963, 10.

80.   *Enquirer* (Battle Creek, MI), December 2, 1948, 33.

81.   Mrs. Eugene Clerisi, "The Discovery of Oil in Hillsdale County," in *150 Years in the Hills and Dales* (Hillsdale, MI: Hillsdale County Historical Society, 1976), 83–84.

82.   *Enquirer* (Battle Creek, MI), January 9, 1957, 22.

83.   *State Journal* (Lansing, MI), July 9, 1961, 33.

84.   *Daily News* (Hillsdale, MI), June 25, 1963, 10.

85.   *Evening Sentinel* (Holland, MI), January 8, 1957, 2; *Daily Globe* (Ironwood, MI), October 1, 1956, 2.

86.   J. B. Buehner and S. H. Davis Jr., "Albion-Pulaski-Scipio-Trend Field," in *Oil and Gas Fields Symposium* (Michigan Basin Geological Society, 1968), 37–41.

87.   Oil companies today use three-dimensional seismic data to find similar fields, but the seismic techniques of the 1950s were not refined enough to detect Albion-Scipio.

88.   Robert E. Ives, "Scipio—Hottest Thing in Michigan," *Oil and Gas*

Journal (September 1958), 214–19.

89. *Gazette* (Cedar Rapids, IA), June 6, 1979, 9B.

90. Ibid., January 28, 1979, 3B.

91. *Daily Herald* (Chicago, IL), August 22, 1979, 2.

92. *Daily Register* (Des Moines, IA), August 25, 1979, 1.

93. David R. Irwin, "Dr. Richard Ireland: Mystic or Magician," *New Times Weekly* (Phoenix, AZ), March 14, 1979, 1.

94. *Arizona Republic*, E1. Ireland made the statement in defense of West's honor as a heterosexual female. Jill Watts, *Mae West: An Icon in Black and White* (New York: Oxford University Press, 2001), 293.

95. Form W-2 for Resource Technology #1 Moore, API #42-353-309459, Texas Railroad Commission files, September 5, 1980.

96. "Richard Ireland Finds Minerals Using His Psychic Ability," video, YouTube, May 17, 2009, https://www.youtube.com/watch?v=lhFXP4opOQ.

97. John R. Shaw Jr., interview by Edmond C. Hardin, December 31, 1963, National Archives, Records of the John F. Kennedy Assassination Collection, Key Persons Files, Ruby, Jack 2-2 Associates and Relatives.

98. A. L. Gary, *The Psychic World of Doc Anderson* (Atlanta: Droke House/Hallux, 1973), 31–34.

99. *Independent* (Gallup, NM), October 13, 1942, 6; Robert E. Smith, *Doc Anderson: The Man Who Sees Tomorrow* (New York: Paperback Library, 1970), 38, 86, 172.

100. *News-Free Press* (Chattanooga, TN), June 28, 1944, 1.

101. Ibid., September 14, 1973, 27.

102. *Reporter-News* (Abilene, TX), July 17, 1969, B1; *Register* (Danville, VA), "Family Weekly" section, December 31, 1967, 4; *Light* (San Antonio, TX), December 28, 1975, 16-A.

103. *Reporter-News* (Abilene, TX), November 23, 1969, 6-A; August 4, 1971, 12-B; David Kaufman, *Doris Day* (New York: Virgin, 1968), 444, 472.

104. *Reporter-News* (Abilene, TX), January 4, 1976, 1; January 6, 1976, 5-A.

105. *Herald* (Titusville, PA), September 26, 1979, 5.

106. *News* (Hutchinson, KS), March 22, 1980, 20.

107. *News-Free Press* (Chattanooga, TN), March 22, 1980, 1; March 23, 1980, 1.

108. *Tribune* (Kokomo, IN), February 23, 1989, 17.

109. D. H. Stormont, "U.S. Will Add Crude and Downstream Capacity This Year," *Oil & Gas Journal* 65, no. 14 (April 3, 1967), 141–43.

110. *Daily Oklahoman* (Oklahoma City, OK), August 25, 1987, 34.

111. Michael Wallis, *Oil Man: The Story of Frank Phillips and the Birth of Phillips Petroleum* (New York: Doubleday, 1988), 109.

112. *Sentinel* (Orlando, FL), September 23, 1972, 4-B; Letter, Harold Sherman to Stan, August 26, 1972, Harold Sherman Collection, University of Central Arkansas, Conway, collection M87-8, series 1, box 9, file 5.

113. *Herald* (Miami, FL), March 18, 1973; *Tropic Magazine*, 16–26.

114. Letter, W. A. Roberts to Ingo Swann, October 25, 1973, MS 8060, Ingo Swann Collection, University of West Georgia, Carrollton, Georgia, box 129, folder 2.

115. *Daily Oklahoman* (Oklahoma City, OK), April 1, 1973, 31.

116. Jim Schnabel, *Remote Viewers* (New York: Dell, 1997), 173; John L. Wilhelm, *The Search for Superman* (New York: Pocket Books, 1976), 201–2.

CHAPTER 5

1. "News and Notes," *American Society of Dowsers Quarterly Digest* 9, no. 1 (February 1969), 6–7; *Evening Standard* (Unionville, PA), August 18, 1962, 7; Jim Anderson, "The Dowsers," *American Dowser* 17, no. 3 (August 1977), 104; "Notices," *Journal of the British Society of Dowsers* 18, no. 191 (March 1981): 1.

2. "News and Notes," *American Society of Dowsers Quarterly Digest* 9, no. 1 (February 1969), 6–7; Albuquerque *Journal*, October 30, 1952, 11; Peter Underwood, *The Complete Book of Dowsing and Divining* (London: Rider, 1980), 91; J. Cecil Maby and T. Bedford Franklin, *The Physics of the Divining Rod* (London: Bell, 1939), 22; "The Buried Treasure Racket," *Popular Mechanics Magazine* 67, no. 2 (February 1937), 162.

3. Jim Kuebelbeck, "L Rods—the Instrument of Choice?" *Dowsing Today* 41, no. 291 (March 2006), 19; Howard V. Chambers, *Dowsing,*

Divining Rods, and Water Witching for the Millions (Los Angeles: Sherbourne, 1969), 57; Kenneth Roberts, *Water Unlimited* (Garden City, NY: Doubleday, 1957), 91.

4. Anne Jaeger-Nosal, *Les Chercheurs D'Eau* (Georg, 1999), 140; William Barrett and Theodore Besterman, *The Divining Rod* (New Hyde Park, NY: University Books, 1968), 13.

5. *Evening Bulletin* (Marysville, KY), February 22, 1888, 1; Gaston Burridge, "Miracles of Map Dowsing," *Fate* 11, no. 6 (June 1958), 71–80; *Daily News* (Galveston, TX), February 20, 1902, 6; *Oil & Gas Journal*, June 21, 1918, 1; July 26, 1934, 74; *State Journal* (Lincoln, NE), September 15, 1927, 5; London *Standard*, January 16, 1912, 10; "Activities of Members," *American Dowser* 13, no. 3 (August 1973), 103; "Doodle Craze Spreading to Mining Areas," *Montana Oil Journal*, September 8, 1928, 2.

6. Benjamin Silliman, "The Divining Rod," *The American Journal of Science and Arts* 11, no. 2 (October 1926): 212.

CHAPTER 6

1. Frances Sullivan, "Don't Call It Magic," *The American Dowser* 15, no. 2 (May 1975), 83–88; Herbert Asbury, *The Golden Flood* (New York: Knopf, 1942), 178–79.

2. *Mining & Petroleum Standard*, January 2, 1865, 193; *American Journal of Mining*, August 17, 1867, 104; *Appleton's Journal*, May 20, 1871, 591; William Wright, *The Oil Regions of Pennsylvania* (New York: Harper Bros., 1865), 62; *The Derrick's Hand-Book of Petroleum* (Oil City, PA: Derrick, 1898), 961.

3. F. W. Minshall, "The History and Development of the Macksburg Oil Field," in *Report of the Geological Survey of Ohio*, vol. 6 (1888), 465.

4. Dallas *Morning News*, August 18, 1940; W. G. Matteson, "The Practical Value of Oil and Gas Bureaus," *Colorado School of Mines Magazine* 7, no. 10 (October 1917), 173; Earl B. Groff, "Letters," *American Dowser* 28, no. 4 (Fall 1988), 68.

5. *Republican* (Van Wert, OH), March 1, 1888, 5; R. R. Brooks, *Biological Methods of Prospecting for Minerals* (New York: Wiley, 1983), 93–97.

6. *Post* (Washington, DC), June 11, 1915, 6.

7. *Darke County Democratic Advocate* (Greenville, OH), January 26, 1888, 1.

8. *Dunn County News* (Menomonie, WI), January 4, 1889, 5.

9. *Daily Tribune* (Chicago, IL), February 29, 1888, 3.

10. N. H. Winchell, *Natural Gas in Minnesota*, Bulletin 5, Geological and Natural History Survey of Minnesota (1889); *Freeborn County Standard* (Albert Lea, MN), August 10, 1887, 4.

11. N. H. Winchell, *Natural Gas in Minnesota*, Bulletin 5, Geological and Natural History Survey of Minnesota (1889).

12. *Daily Tribune* (Chicago, IL), May 26, 1888, 1.

13. *Daily Democratic Times* (Lima, OH), February 22, 1888, 4; *Crawford County Forum* (Bucyrus, OH), January 6, 1888, 5.

14. A. da Silva Mello, *Mysteries and Realities of This World and the Next*, trans. M. B. Fierz (London: Wiedenfeld & Nicolson, 1960), 149.

15. Max Dessoir, "Ein Hellseher," *Psychische Studien* 34, no. 9 (September 1907): 567.

16. Salt Lake (UT) *Tribune*, May 10, 1891, 8; Steubenville (OH) *Daily Herald*, December 27, 1888, 2; Muncie (IN) *Daily Herald*, October 1, 1890, 3; Logansport (IN) *Journal*, April 8, 1892, 5; Scranton (PA) *Republican*, November 14, 1910, 3.

17. Elizabeth Loftus and Katherine Ketcham, *Witness for the Defense* (New York: St. Martin's, 1991); Theodore Anneman, *Practical Mental Magic* (1944; rpt. New York: Dover, 1983), 7–11.

18. Martin Gardner, *Did Adam and Eve Have Navels?* (New York: Norton, 2000), 217; Eugene Osty, *Supernormal Facilities in Man* (London: Methuen, 1923), 27–29; Joseph F. Rinn, *Sixty Years of Psychic Research* (San Diego, CA: Truth Seeker Co., 1950), 142–47; Bailey Millard, "What Is There in the Occult?" *Illustrated World* 24, no. 5 (January 1915), 631–32.

19. Nandor Fodor, *An Encyclopedia of Psychic Science* (Secaucus, NJ: Citadel, 1966), 325; Massimo Polidoro, "The Man Who fooled Edison . . . but not Houdini," *Skeptical Inquirer* 31, no. 5 (September/October 2007), 25.

20. Eugene Osty, *Supernormal Faculties in Man*, trans. Stanley de Brath (London: Methuen, 1923), 27.

21. Alois Wiesinger, *Occult Phenomena in the Light of Theology*, trans.

Brian Battershaw (London: Burns & Oates, 1957), 198.

22. Gerald T. White, *Formative Years in the Far West* (New York: Appleton-Century-Crofts, 1962), 195; Ron Chernow, *Titan* (New York: Random House, 1998), 77, 132, 284.

23. *Herald* (Titusville, PA), April 19, 1895, 2.

24. *Express* (Fort Collins, CO), February 26, 1902, 8.

25. Ibid.

26. Donald H. Kupfer, "Cañon City's Oil Spring, Fremont County, Colorado: Colorado's First Commercial Oil Prospect (1860); and the Discovery of the Florence Oil Field (1881)," *Oil-Industry History* 1, no. 1 (2000), 35–59; Chester W. Washburne, "The Florence Oil Field, Colorado," in US Geological Survey, Bulletin 381 (1909) gives the Florence field a discovery date of 1876. Kupfer has shown this to be in error ["Discovery Date of Florence Oil Field," *The Outcrop: Newsletter of the Rocky Mountain Association of Geologists* (Denver, CO: June 1998), 5].

27. George H. Eldridge, "The Florence Oil-field, Colorado," *Transactions of the American Institute of Mining Engineers* 20 (June–October 1891), 442–62.

28. *Express* (Fort Collins, CO), September 22, 1881, 4; Clarence W. Washburne, "The Florence Oil Field, Colorado," in *Contributions to Economic Geology*, US Geological Survey, Bulletin 381, 533.

29. *Mail* (Salida, CO), December 6, 1892, 6.

30. *Derrick* (Oil City, PA), April 18, 1895, 4.

31. *Daily Camera* (Boulder, CO), February 27, 1902, 1.

32. N. M. Fenneman, *Geology of the Boulder District, Colorado*, US Geological Survey, Bulletin 265 (1905).

33. *Daily Camera* (Boulder, CO), February 21, 1902, 1.

34. *Express* (Fort Collins, CO), September 24, 1902, 10.

35. *Evening Tribune* (Marysville, OH), November 6, 1902, 2.

36. *Baylor County Banner* (Seymour, TX), April 3, 1919, 1.

37. *Daily Record* (Roswell, NM), March 5, 1906, 3; June 11, 1906, 1; State of New Mexico Oil Conservation Division database, Canfield State #001, API # 20-005-01230.

38. *Daily Record* (Roswell, NM), June 1, 1907, 4; Phil D. Helmig, "History of Petroleum Exploration in Southeast New Mexico," *The*

Oil and Gas Fields of Southeastern New Mexico, eds. T. F. Stipp et al. (Roswell, NM: Roswell Geological Society, 1956), 21–26.

39. *Daily Record* (Roswell, NM), May 9, 1907, 4.

40. *Oil, Paint, and Drug Reporter*, August 11, 1919, section II, 23.

41. *Daily Phoenix* (Muskogee, OK), April 26, 1915, 15.

42. "Southwest Texas," *Oil & Gas Journal* 18, no. 12 (August 22, 1919), 20.

43. "Personal Mention—Men You Know," *Oil Weekly* 35, no. 8 (November 14, 1924), 29.

CHAPTER 7

1. Harold F. Williamson and Arnold R. Daum, *The American Petroleum Industry, vol. 1: The Age of Illumination 1859–1899* (Evanston, IL: Northwestern University Press, 1959), 90.

2. Piotr Krzywiec, "The Birth and Development of the Oil and Gas Industry in the Northern Carpathians (up until 1939)," *History of the European Oil and Gas Industry*, eds. Craig et al., Special Publication 465 (London: Geological Society, 2018), 179; Edgar Wesley Owen, *Trek of the Oil Finders: A History of Exploration for Petroleum* (Tulsa, OK: American Association of Petroleum Geologists, 1975), 1355.

3. Monika Gisler, "Entangled between Worlds: Swiss Petroleum Geologists, c. 1900–50," *History of the European Oil and Gas Industry*, eds. Craig et al., Special Publication 465 (London: Geological Society, 2018), 369; Wilbur E. McMurtry and Edgar W. Owens, "Everett Carpenter—Historical Notes," *American Association of Petroleum Geologists Bulletin* 52, no. 9 (September 1968), 1800–1803; *Mexico's Oil* (Mexico City: Government of Mexico, 1940), 12–13.

4. Bob Nelson, *Old Town Orcutt* (Orcutt, CA: Orcutt Historical Committee, 1987), 25; J. R. Pemberton, "Economics of the Oil and Gas Industry of California," *Geologic Formations and Economic Development of the Oil and Gas Industry of California*, Olaf P. Jenkins, California Division of Mines, Bulletin 118 (April 1943), 6; Michel T. Halbouty, "Giant Oil Fields in the United States," *Geology of Giant Petroleum Fields*, ed. Michel T. Halbouty (Tulsa, OK: American Association of Petroleum Geologists, 1970), 91–127.

5. Ida M. Tarbell, Introduction to Paul H. Giddens, *The Birth of the Oil*

Industry (New York: MacMillan, 1938), xxxix.

6.   Ellen Sue Blakey, *Oil on Their Shoes* (Tulsa, OK: American Association of Petroleum Geologists, 1985), 75.

7.   "Production and daily average each month of all the Oklahoma districts excepting Healdton in 1915," *Fuel Oil Journal* 7, no. 2 (February 1916), 5.

8.   "New Publications," *Engineering & Mining Journal* 26, no. 12 (September 21, 1878), 200.

9.   Wilbur E. McMurtry and Edgar W. Owen, "Everett Carpenter— Historical Notes," *American Association of Petroleum Geologists Bulletin* 52, no. 9 (September 1968), 1800–1803.

10.   Wallace Pratt, "Oil Finding—the Way It Was," *Oil & Gas Journal* 75, no. 35 (August 1977), 143–45; Ellen Sue Blakey, *Oil on Their Shoes* (Tulsa, OK: American Association of Petroleum Geologists, 1985), 99.

11.   James H. Gardner, "Work of Modern Geologist," *Oil Weekly* 44, no. 14 (November 26, 1914), 3; Wallace E. Pratt, "Geology Came to Lead Industry and Stayed to Learn," *Oil Weekly* 44, no. 13 (March 18, 1927), 37–42; H. C. Fowler, *Developments in the American Petroleum Industry, 1914–1919*, US Bureau of Mines, Information Circular 7171, June 1941, 6.

12.   Joseph A. Kornfeld, "A Half Century of Exploration," *Oil & Gas Journal* (May 1951), 198; Ellen Sue Blakey, *Oil on Their Shoes* (Tulsa, OK: American Association of Petroleum Geologists, 1985), 99; *Oil & Gas Journal*, August 26, 1918, 3; W. A. Ver Wiebe, *North American and Middle East Oil Fields* (Wichita, KS: self-published, 1950), 73.

13.   Edgar Wesley Owen, *The Trek of the Oil Finders: A History of Exploration for Petroleum*, Introduction by Wallace Pratt (Tulsa, OK: American Association of Petroleum Geologists, 1975), xiii–xv.

14.   Lloyd E. Gatewood, "Arbuckle Environments: Some Models and Examples," *Shale Shaker Digest* 9 (1979), 34–45; Sam S. Josephson, "Montana Oil Fields," *Petroleum* 11, no. 2 (June 1921), 126.

15.   US Geological Survey, *Mineral Resources of the United States*, annual reports for years 1908 to 1923; US Bureau of Mines, *Mineral Resources of the United States*, annual reports for years 1925 to 1976; "North American Drilling Activity," *American Association of Petroleum Geologists Bulletin*, annual series of articles for years 1977 to 1989.

16. *Daily News* (Denver, CO), November 9, 1902, 2; *New York Times*, November 16, 1902, 33.

17. *Herald* (El Paso, TX), May 5, 1903, 1.

18. *Daily News* (Galveston, TX), March 15, 1905, 22.

19. *Daily Review* (Bisbee, AZ), July 10, 1907, 6.

20. *Express* (San Antonio, TX), October 26, 1936, 8; *Oil & Gas Journal*, November 8, 1917, 36.

21. *Evening News* (Modesto, CA), June 22, 1922, 4; *Wichita Daily Times* (Wichita Falls, TX), August 15, 1915, 7.

22. *News* (Hutchinson, KS), January 19, 1924, 8; *Daily Eagle* (Wichita, KS), January 4, 1918, 10; "Let Henry Zachary," *Oil & Gas Journal*, April 26, 1917, 39.

23. *Post-Register* (Lockhart, TX), March 18, 1920, 17.

24. Robert L. Morlan, "Townley, Arthur Charles," *Dictionary of American Biography, Supplement Six*, ed. John A. Garraty (New York: Scribner's, 1980), 644–45.

25. Robert L. Morlan, *Political Prairie Fire* (Minneapolis: University of Minnesota Press, 1955), 23–27.

26. *Times* (Reading, PA), August 25, 1924, 1; *Evening Gazette* (Cedar Rapids, IA), October 24, 1921, 13.

27. John Bluemle, "Prospects and Swindles," *Rocky Mountain Oil Journal*, August 30, 1996, 3.

28. *Tribune* (Bismarck, ND), August 14, 1925, 1.

29. Ibid., August 23, 1926, 1; May 14, 1927, 1; May 16, 1927, 1.

30. Ibid., December 28, 1929, 3; August 2, 1927, 1.

31. Ibid., May 29, 1952, 4.

32. Ibid., June 5, 1952, 12.

33. *Press* (Pittsburgh, PA), June 13, 1952, 25; *Tribune* (Great Falls, MT), June 5, 1952, 4; *Tribune* (Bismarck, ND), June 3, 1952, 1.

34. Robert L. Morlan, "Townley, Arthur Charles," *Dictionary of American Biography, Supplement Six*, ed. John A. Garraty (New York: Scribner's, 1980), 644–45.

35. *Star* (Indianapolis, IN), September 24, 1922, 8; advertisement, *Literary Digest* 48, no. 11 (March 14, 1914), 569; "The Propaganda for Reform: Swoboda's 'Conscious Evolution,'" *Journal of the American Medical Association* 70, no. 11 (March 16, 1918): 799–802.

36. *News* (Hutchinson, KS), July 2, 1924, 4.

37. "Mad or Bad?" *Engineering and Mining Journal* 118, no. 26 (December 27, 1924), 1002.

38. George H. Hansen and H. C. Scoville, *Drilling Records for Oil and Gas in Utah*, Utah Geological and Mineralogical Survey, Bulletin 50 (1955), 18; *Telegram* (Salt Lake City, UT), November 8, 1928, 5; *Herald* (Burley, ID), September 11, 1928, 1.

39. *Tribune* (Bismarck, ND), April 15, 1930, 1.

40. New York *Daily News*, June 27, 1930, 19.

41. L. V. Pirsson and T. Wayland Vaughan, "A Deep Boring in Bermuda Island," *American Journal of Science* 186 (1913), 4th Series, 70–71; Kenneth Roberts, *Henry Gross and His Dowsing Rod* (Garden City, NY: Doubleday, 1951), 163–65, 184; Kenneth Roberts, *The Seventh Sense* (Garden City, NY: Doubleday, 1953), 65; *Courier-Journal* (Louisville, KY), January 15, 1951, 4; T. E. Coalson II, "Dowsing, the Eternal Paradox," *Psychic* 5, no. 4 (March/April 1974), 14; Richard Wolkomir, "Water Witches," *Omni* 7, no. 12 (September 1995), 42.

42. James A. M. Thompson, "Modeling Ground-water Management Options for Small Limestone Islands: The Bermuda Example," *Ground Water* 27, no. 2 (March/April 1989), 147–54.

43. John W. McDonald, with Noa Zanolli, *The Shifting Grounds of Conflict and Peacebuilding* (London: Rowman and Littlefield, 2008), 54; *Journal* (Biddleford-Saco, ME), February 18, 1959, 1; Kenneth Roberts, *Water Unlimited* (Garden City, NY: Doubleday, 1957), 59, 144–51, 208–14, 224–30; C. R. Roseberry, "Gross and Groundwater," *Water Well Journal* (March 1955), 12–54; Kenneth Roberts, *The Seventh Sense* (Garden City, NY: Doubleday, 1953), 244–51; Kenneth Roberts letter to John Masterson, May 15, 1957, box 24, folder 71, Kenneth Roberts Collection, Dartmouth College Library, Hanover, New Hampshire; Edwin M. Shook, *Incidents in the Life of a Maya Archaeologist* (Guatemala City: Associación de Amigos del Pais y Fundación para la Cultura y el Desarollo, 1998), 127–28.

44. Kenneth Roberts, *Water Unlimited* (Garden City, NY: Doubleday, 1957), 223.

45. Robert L. Williams letter to Kenneth Roberts, April 9, 1956, box 24, folder 70, Kenneth Roberts Collection, Dartmouth College Library,

Hanover, New Hampshire.

46. The Texas sharpshooter fallacy imagines that a shooter fires bullets into the side of a barn, then draws a bullseye around each bullet hole and brags of his skill in hitting each bullseye.

47. Berthold E. Schwarz, *A Psychiatrist Looks at ESP* (New York: New American Library, 1965), 116–17. The well is the Egbert #6 (API #16-025-00262), which Schwarz calls #9. Well files of Kentucky Geological Survey.

48. The deep wells are the #1 Egbert (API #16-025-00263) and #2 Egbert (#16-025-00265).

49. "Obituary," *American Dowser* 19, no. 4 (November 1979), 148; Francis Hitching, *Dowsing* (Garden City, NY: Doubleday, 1978), 55.

50. *News-Journal* (Mansfield, OH), January 31, 1960, 58.

51. *Star* (Marion, OH), January 12, 1960, 1.

52. *News-Journal* (Mansfield, OH), August 5, 1973, 6-A.

53. W. E. Shafer, "Historic Impressions, Seismic Observations; The Evolution of a Geologic Model and Other Comments," *The Ohio Geological Society Anthology on Morrow County, Ohio "Oil Boom" 1961–1967 and the Cambro-Ordovician Reservoir of Central Ohio*, ed. William E. Shafer (Columbus: Ohio Geological Society, 1994), 3–32; *News Journal* (Mansfield, OH), April 16, 1964, 15.

54. Christopher Bird, "His Name Is Dr. Drill," *American Dowser* 20, no. 4 (November 1980), 28–29.

55. Christopher Bird, *The Divining Hand* (New York: Dutton, 1979), 185, 192, 197.

56. James E. Coggin, *J. K. Wadley: A Tree God Planted* (Texarkana, AR: Southwest Printers, 1971), 40–41, 59–60.

57. Laile E. Bartlett, *Psi Trek* (New York: McGraw-Hill, 1981), 19.

58. John Fairley and Simon Welfare, *Arthur C. Clarke's World of Strange Powers* (London: Collins, 1989), 186.

59. Christopher Bird, *The Divining Hand* (New York: Dutton, 1979), 181, 185–90.

60. Email from Jimmie Joe Ault to Dan Plazak, April 8, 2015.

61. Ibid., and April 9, 2015.

62. Christopher Bird, *The Divining Hand* (New York: Dutton, 1979), 189, 192–96. Bird called the mineral *laurelite* but evidently meant *laurite*.

63. Earl Pyle, "From Water to Oil," *American Society of Dowsers Quarterly Digest* 5, no. 3 (May 1965), 34–35; Earl Pyle, *How to Make a Million Dowsing and Drilling for Oil* (Hicksville, NY: Exposition, 1977), 136.

64. *Courier-Journal* (Louisville, KY), August 27, 1959, 5.

65. "Activities of Members," *American Dowser* 13, no. 3 (August 1973), 103; Earl Pyle, "Letters to the Editor," *American Dowser* 13, no. 1 (February 1973), 23.

66. *Advocate-Messenger* (Danville, KY), April 7, 1986, 3.

67. John W. Jewell, *Tales of the Third World and Appalachia* (n.l.: self-published, 1999), 104–5.

68. *Courier-Journal* (Louisville, KY), February 22, 1988, 3.

69. *Park City Daily News* (Bowling Green, KY), May 21, 1991, 5.

70. William Trombley, "The Dowser Who Finds Oil Wells," *Saturday Evening Post* 236, no. 43 (December 7, 1963), 88–89.

71. Nelson #1 Breckbill, API 15-041-00045.

72. *Register* (Des Moines, IA), January 27, 1974, 1.

73. *Journal* (Salina, KS), May 20, 1998, 7.

74. *Tribune* (South Bend, IN), December 1, 1966, 1.

75. Santa Fe *New Mexican*, April 6, 1980, 4.

76. *Florida Today* (Cocoa, FL), March 9, 1980, 10B; *Democrat* (Tallahassee, FL), 1.

77. L. F. Willard, "When Water Witches Get Together," *Yankee* (September 1979), 28–43; London *Daily Mirror*, September 27, 1973, 25.

78. *Tribune* (South Bend, IN), December 1, 1966, 1; *Daily Times-Mail* (Bedford, IN), February 17, 1974, 15; December 3, 1978, 43.

79. *Tribune* (South Bend, IN), September 8, 1991, B12.

80. *Daily Leader* (Marietta, OH), June 16, 1897, 3; "Moosman, the West Virginia Oil 'Smeller,' Meets His Death—His Power Baffled Science," *Paint, Oil and Drug Review* 23, no. 25 (June 23, 1897), 24.

81. *Enquirer* (Cincinnati, OH), June 26, 1897, 14.

82. *Banner* (Bristol, IN), August 14, 1914, 7; *Telegram* (Hartford City, IN), May 2, 1900, 4; *Daily Advocate* (Newark, OH), December 24, 1897, 7.

83. *Star-Press* (Muncie, IN), October 12, 1907, 6; *Evening Journal* (Hamilton, OH), June 17, 1892, 3.

84. *Hamilton County Ledger* (Noblesville, IN), July 4, 1899, 1.

85. *Daily News* (Arkansas City, KS), December 30, 1913, 1.

86. *Tribune* (Arkansas City, KS), May 16, 1929, 8; *Daily News* (Arkansas City, KS), February 2, 1916, 5; *Eagle* (Burden, KS), April 9, 1903, 5; *Daily Traveler* (Arkansas City, KS), December 4, 1902, 5.

CHAPTER 8

1. R. B. Harkness, "Gas and Oil in Eastern Ontario," *46th Annual Report of the Ontario Department of Mines, 1937* 46, part 5 (1938), 101–6.

2. R. B. Harkness, "Petroleum in 1927," *37th Annual Report of the Ontario Department of Mines*, part 5 (1928), 44–50.

3. "Etheridge Making More Lucky Strikes," *Western Oil Examiner* 3, no. 25 (August 4, 1928), 3. *Herald* (Calgary, Alberta), March 20, 1939, 19.

4. *Herald* (Lethbridge, Alberta), March 31, 1951, 8; *Republican* (Council Grove, KS), February 13, 1951, 1; John O. Galloway, "Developments in Western Canada in 1950," *Bulletin of the American Association of Petroleum Geologists* 35, no. 6 (June 1951), 1395; A. K. Chetin, "Barons Field," *Oil Fields of Alberta* (Calgary, Alberta: Alberta Society of Petroleum Geologists, 1960), 60–61.

5. *Journal* (Edmonton, Alberta), June 12, 1980, 5; Calgary, *Albertan*, March 22, 1962, 26.

6. Kurt Reden, "Madame 'Charlotte von Tukory,' the 'Wonder' Woman," *Petroleum Times* 13, no. 327 (April 11, 1925), 682.

7. Anne Jaeger-Nosal, *Les Chercheurs d'Eau* (Geneva: Georg, 1999), 20–21.

8. J. Durand, "La Recherche du Pétrole en France," *Houille, Minerais, Pétrole* 1, no. 1 (January–February 1946): 10.

9. Pierre-Louis Maubeuge, *Comme une Odeur de Pétrole* (Pierron, 1996), 27, 189.

10. Louis Barrabé, "Radiesthésie et Prospection Geologique," *La Radiesthésie: Études Critiques* (Paris: L'Union Rationaliste, 1956), 35–48.

11. J. Durand, "Sur la Sourcellerie," *Houille, Minerais, Pétrole* 1, no. 2 (March–April 1946): 47–48; A. Roux and M. Solignac, "Historique et État Actuel des Recherches de Pétrole en Tunisie," *IIme Congrès Mondiale du Pétrole* (Paris, 1937), 213–29.

12. Monica Gisler, "Entangled Between Two Worlds: Swiss Petroleum Geologists, c. 1900–50," *History of the European Oil and Gas Industry*, ed. J. Craig et al., Special Publication 465 (London: Geological Society, 2018), 361–74; Marc Weidmann, "Histoire de la Prospection de l'Exploitation des Hydrocarbures en Pays Vaudois," *Bulletin de la Société Vaudoise des Sciences Naturelles* 80, no. 4 (December 1991), 365–402.

13. Patrick H. LaHusen and Roland Wyss, "Erdöl und Erdgasexploration in der Schweiz: Ein Rückblick," *Bulletin Schweiz Verein des Petroleum Geologen und Ingenieuren* 62, no. 141 (December 1995): 43–72.

14. R. F. Rutsch, "Switzerland," *Bulletin of the American Association of Petroleum Geologists* 37, no. 7 (July 1953): 1646–647.

15. Arnold Heim, "Wünschelrute im Schweiz," *Zeitschrift für Angewandte Geologie* 3, no. 8/9 (1957): 426–27.

16. *Evening News* (Harrisburg, PA), April 18, 1929, 14.

17. *Evening Post* (Nottingham, England), June 29, 1932, 6.

18. Marc Wiedmann, "Histoire de la Prospection et de l'Exploitation des Hydrocarbures en Pays Vaudois," *Bulletin de Géologie Lausanne*, no. 314 (1991): 365–402; Abbé Mermet, *Principles and Practice of Radiesthesia*, trans. Mark Clement (London: Watkins, 1959), 87.

19. M. F. Cazzamali, "Rhabdomancie," *Revue Métapsychique*, no. 5 (September–October 1931): 337–55; *Daily Globe* (Boston, MA), October 26, 1922, 11.

20. Nicholas Farrell, *Mussolini: A New Life* (London: Phoenix, 2003), 323; Francobaldo Chiocci, *Donna Rachele* (Rome: Ciarrapico, 1983), 250–51.

21. Rudolph Delkeskamp, "Fortschritte auf dem Gebiete der Enforschung der Mineralquellen," *Zeischrift für Praktische Geologie* (August 1908): 27–443.

22. Donald C. Barton, "The Wigglestick," *Bulletin of the American Association of Petroleum Geologists* 7, no. 4 (July 1923), 427–29.

23. Karl Engler, *Das Erdöl* (Verlag, 1930), 4: 537.

24. Wilhelm Meseck, "Rutengänger, Erdöl und Wissenschaft," *Zeitschrift für Radiästhesie* (July 1966), 18, Jahrgang n.3, 91–102; Carl Streich, "The New Oilfields of Germany," *Petroleum Times* 14, no. 359 (November 21, 1925), 876.

25. *Journal of the British Society of Dowsers* 3, no. 24 (June 1939), 387; Karl von Klinckowstrom, "The Present Position of the Divining Rod Question in Germany," *Journal of the Society for Psychical Research* 22, no. 414 (April 1925): 54–60; Donald C. Barton, "The Wigglestick," *Bulletin of the American Association of Petroleum Geologists* 10, no. 3 (March 1926): 312–13.

26. R. B. Behrmann, "Geophysical History of the Reitbrook Salt Dome and Oil Field near Hamburg, Northwest Germany," *Geophysical Case Histories*, ed. L. L. Nettleton, vol. 1 (Tulsa, OK: Society of Exploration Geophysicists, 1949), 619–26.

27. Carl Streich, "The New Oilfields of Germany," *Petroleum Times* 14, no. 359 (November 21, 1925); "The Development of North German Oil Territory," *Petroleum Times* 23 (May 11, 1930), 961–62; P. Zurcher, "European Opinion of the Divining Rod," *Engineering & Mining Journal* (January 5, 1929), 20.

28. "Difficulty of Keeping Abreast of Progress," *Petroleum Times* 15, no. 374 (March 6, 1926), 402.

29. "The Development of North German Oil Territory," *Petroleum Times* 23 (May 31, 1930), 961–62.

30. "The Use of the Divining Rod," *The Surveyor and Municipal County Engineer* 30, no. 755 (July 6, 1906), 25.

31. Schubert, "Beglaubigte Abschrift," *Schriften der Verbands zur Klärung der Wünschelrute* 7 (1916), 17–24.

32. *New York Tribune*, July 3, 1914, 1; *Daily News* (Passaic, NJ), October 17, 1914, 1.

33. Butte *Montana Standard*, March 19, 1915, 10; Martin Nordegg, *The Possibilities of Canada Are Truly Great* (Toronto: MacMillan, 1971), 204–5.

34. Otto Prokop, *Wünschelrute Erdstrahlen und Wissenschaft* (Stuttgart: Verlag, 1955), 162.

35. "The Divining Rod in Austria," *Mining Journal* 105 (June 20, 1914), 588.

36. "New Oil Drilling in Austria," *Petroleum World* 22, no. 293 (February 1925), 74.

37. Hans Falkinger, "Zistersdorf und die Wünschelrute," *Zeitschrift für Wünschelrutenforschung* 20, no. 7/8 (July/August 1939): 137–41.

38. Leicester (England) *Illustrated Chronicle*, October 10, 1936, 5.

39. *Petroleum World* 23, no. 319 (July 1926), 264; Emerich Herzog, *Wünschelrute* (Vienna: Stenrermuhl-Verlag, 1926), 20; *Jewish News* (Denver, CO), May 3, 1922, 1.

40. London *Guardian*, September 26, 1936, 12; Liverpool *Echo*, August 27, 1936, 12.

41. George Eshelby, "Another Way of Finding Oil in Britain," *Petroleum Times* 38, no. 978 (October 9, 1937), 461.

42. Devon *Western Morning News*, March 8, 1938, 4; *Gleaner* (Kingston, Jamaica), October 19, 1936, 21.

43. F. L. M. Boothby, "Can the Dowser Assist British Oil Search?" *Petroleum Times* 38, no. 966 (July 17, 1937), 80; "Oil in Southern England?" *Petroleum Times* 39, no. 990 (January 1, 1938), 17–18.

44. E. H. Cunningham-Craig, "Petroleum in Scotland?" *Petroleum Times* 36, no. 926 (October 10, 1936), 455–56.

45. "Oil Prospecting in England," *Chemical Age* 35, no. 903 (October 17, 1936), 337.

46. "Water Diviners of India," *Journal of the British Society of Dowsers* 5, no. 40 (June 1943), 222–30.

47. A. H. Bell, "A Claim for the Oil Dowser," *Petroleum Times* 40 (August 20, 1938), 214; "British Oil Search—Dowsing the Dowser," *Petroleum Times* 40 (September 3, 1938), 296; A. H. Bell, "Dowsing for Oil," *The Spectator*, no. 5747 (August 19, 1938), 306; "'Shooting' Anglo-American Oil Co.'s Second Producing Well in Scotland," *Petroleum Times* 42 (July 22, 1939), 124–25.

48. Evelyn Penrose, *Adventure Unlimited* (London: Spearman, 1958), 11, 48, 94–95.

49. Victoria (British Columbia) *Times Colonist*, April 25, 1931, 15.

50. *Western Mail* (Perth, West Australia), November 11, 1954, 3.

51. Evelyn Penrose, *Adventure Unlimited* (London: Spearman, 1958), 186.

52. Walter Kempthorne, "Evelyn Penrose's Lifetime of Dowsing," *Fate* 22, no. 6 (June 1969), 76–83.

53. Gene Z. Hanrahan, *The Bad Yankee*, vol. 1 (Chapel Hill, NC: Documentary Publications, 1985), 22.

54. Nyls Gustavo Ponce Seoane, "La Radiestesia en Cuba," *Boletín de la Sociedad Cubana de Geología* 6, no. 3 (September–December 2006):

10–14; Ruth Sheldon Knowles, "Cuba, a Challenge to Wildcatters," *World Petroleum* 23, no. 3 (March 1952), 52–55.

55. Semyon Semyonov, "Who Last Looked in the Mirror? The Siberian Folk Psychic Who Could Tell," *The ESP Papers*, eds. Sheila Ostrander and Lynn Schroeder (New York: Bantam, 1976), 131.

56. Boris Goldovsky, "What Psychics Can Do," *Asia and Africa Today* (January 1992), 90–91.

57. N. N. Sochevanov and V. S. Matveev, "Biophysical Method in Geological Research," *Geologia Rudnykh Mestorozhdenii* (Geology of Ore Deposits) 16, no. 5 (July–Aug. 1974): 77–85; N. G. Schmidt, A. N. Eremeev, and A. P. Solovkin, "Is There a Biophysical Method for Searching for Ore Deposits?" *Geologia Rudnykh Mestorozhdenii* (Geology of Ore Deposits) 17, no. 5 (September–October 1975): 88–95.

58. Christopher Bird, "Documentation: Some Soviet Geologists Back Use of Dowsing to Prospect for Natural Resources," *American Dowser* 19, no. 1 (February 1979), 31–32.

59. V. Yermolayev, "With Divining Rod Atilt," *Pravda* (July 5, 1982), condensed English translation in *The Current Digest of the Soviet Press* 34, no. 27 (August 4, 1982), 20; Christopher Bird, "Soviets Take Dowsing Seriously," *Fate* (June 1983), 58–64; *Paraphysics R&D–Warsaw Pact, US Defense Intelligence Agency*, DST-1810S-202-78 (February 4, 1980), 22.

60. V. K. Kozyanin, "Classification and Forecasting of Hydrocarbon Deposits in the Kolpakov Trough of Western Kamchatka," *Geology of the Pacific Ocean* 7, no. 5 (1992), 1191–97 (translation of article from *Tikhookeanskaya Geologiya* 8, no. 5 (1990)).

61. V. Nalivkin, "Clairvoyance and Oil," *Science in Russia*, no. 2/3 (March–June 1993), 94–97.

62. M. Ya. Borovsky et al., "Mobile Geophysics for Searching and Exploration of Domanic Hydrocarbon Deposits," *IOP Conference Series: Earth and Environmental Science* 155 (2018), 1–6; M. Sh. Mardanov, V. N. Dyachkov, and V. B. Podavalow, "The Application of Biogeophysical Studies in the Search for Oil Fields," *Georesursy = Georesources* 19, no. 3, part 2 (2017), 284–91.

63. G. Gabrielyants and V. Poroskun, "The History of Scientific

Foundations of Oil Exploration (Battle of Ideas, Theories and Concepts)," 42nd International Commission on the History of Geological Sciences (INHIGEO) Symposium, Yerevan, Armenia (2017), 113–14; A. Mychak and A. Shybetska, "Comprehensive Approach in Definition of the Oil and Gas Exploration Objects Priority," *Ukrainian Journal of Remote Sensing* 25 (2020): 28–32.

CHAPTER 9

1.   J. A. Simpson and E. S. C. Weiner, "Doodle," *Oxford English Dictionary*, 2nd ed., 4 vols. (Oxford: Clarendon, 1989).

2.   Bob Tippee, "Sticks and Stones," *Oil & Gas Journal* 117, no. 8 (August 5, 2019), 18.

3.   Yes, the line is stolen from Owen Wister's *The Virginian*. But it's still good advice.

4.   *Mirror* (Warren, PA), November 2, 1887, 3; A. R. Crum and A. S. Dungan, *Romance of American Petroleum and Gas* (Romance of American Petroleum and Gas Co., 1911), 198; John J. McLaurin, *Sketches in Crude Oil* (Franklin, PA: self-published, 1902), 108.

5.   *Statesman Journal* (Salem, OR), July 23, 1904, 4; *Daily Capital Journal* (Salem, OR), May 29, 1902, 4.

6.   *St. John's Review* (Portland, OR), April 9, 1909, 1.

7.   Mary Randolph Wilson Kelsey and Mavis Parrott Kelsey Sr., *James George Thompson, 1803–1879* (College Station: Texas A&M University Press, 1988), 511–12.

8.   Ellis A. Davis and Edwin H. Grobe, *The New Encyclopedia of Texas*, 2 vols. (Dallas: Texas Development Bureau, 1925), 1582.

9.   Albert G. Wolf, "Big Hill Salt Dome, Matagorda County, Texas," *Bulletin of the American Association of Petroleum Geologists* 9, no. 4 (July 1925), 711–37.

10. Anthony F. Lucas, "Possible Existence of Deep-Seated Oil Deposits on the Gulf Coast," *Transactions of the American Institute of Mining and Metallurgical Engineers* 61 (1920), 501–16.

11. David T. Day, "Petroleum," *Mineral Resources of the United States 1913*, part II, US Geological Survey (1914), 929–1047.

12. John D. Northrop, "Petroleum," *Mineral Resources of the United States 1914*, part II, US Geological Survey (1916), 893–1023.

13. *Post* (Houston, TX), June 15, 1913, 34.

14. Alexander Deussen, "The Humble (Texas) Oil Field," *Bulletin of the American Association of Petroleum Geologists* 1, no. 1 (January 1917), 60–84.

15. Paul Dickson, *The Official Rules* (New York: Dell, 1978), 71.

16. W. A. Owens, interview, Oral History of the Texas Oil Industry Collection, Dolph Briscoe Center for American History, University of Texas at Austin, July 30, 1953, tape 123; "Voodooism Applied to Mining," *Engineering and Mining Journal* 111, no. 17 (April 23, 1921).

17. *Tribune* (Oakland, CA), November 21, 1915, 19.

18. W. M. Hudson, interview, Oral History of the Texas Oil Industry Collection, Dolph Briscoe Center for American History, The University of Texas at Austin, September 18, 1952, tape 81.

19. Mody C. Boatright, *Folklore of the Oil Industry* (Dallas: Southern Methodist University Press, 1963), 44.

20. *Beacon* (Wichita, KS), July 13, 1914, 2.

21. *Beacon* (Wichita, KS), August 1, 1914, 6.

22. *Daily Eagle* (Wichita, KS), July 31, 1914, 5; *Sun-Monitor* (Magnum, OK), August 27, 1908, 1.

23. *Democrat* (Liberal, KS), May 29, 1914, 6.

24. Altus *Oklahoma Democrat*, April 15, 1915, 1.

25. *News-Democrat* (Elk City, OK), January 13, 1921, 1; *Beacon* (Wichita, KS), July 13, 1914; "Is Oil Geology Menaced?" *Oil & Gas Journal*, May 20, 1921, 3.

CHAPTER 10

1. *Beacon* (Wichita, KS), November 11, 1916, 5.

2. Reno *Daily Nevada State Journal*, June 15, 1892, 1.

3. "Putting One Over on the Rock Hounds," *Oil and Gas News* 5, no. 13 (May 15, 1919), 11.

4. *Spokesman-Review* (Spokane, WA), September 12, 1920, 4; G. R. Hopkins and A. B. Coons, "Petroleum," *Mineral Resources of the United States, 1928*, part II, ed. Frank J. Katz, US Bureau of Mines (1930), 689.

5. Robert Conot, *A Streak of Luck* (New York: Seaview, 1979), 133.

6. F. L. Aurin, "Regarding Fake Geologists," *Marland News* (Ponca City,

OK), November 3, 1920, 3; Lois Osburn, "Oil Finders," unpublished manuscript, Works Progress Administration, Federal Writers' Project, Library of Congress, Mss55715, box A736, November 5, 1936, 8, https://www.loc.gov/item/wpalh002531.

7. "Albert Abrams," *Journal of the American Medical Association* 78, no. 14 (April 8, 1922): 1072.

8. *Herald* (El Paso, TX), May 9, 1923, 1: Upton Sinclair, "The House of Wonder," *Pearson's Magazine* 48, no. 6 (June 1922), 9–17; Austin C. Lescarboura, "Our Abrams Verdict," *Scientific American* 131, no. 9 (September 1924), 158; Ernest W. Page, "Portrait of a Quack," *Hygeia* 17, no. 1 (January 1939): 92.

9. "The Electronic Reactions of Abrams," *Boston Medical and Surgical Journal* 187, no. 7 (August 17, 1922): 268; Austin C. Lescarboura, "Our Abrams Verdict," *Scientific American* 131, no. 9 (September 1924), 158–60, 220–22.

10. "Abrams' Divining Rod: The Apotheosis of Buncombe," *Journal of the American Medical Association* 80, no. 9 (March 3, 1923): 631–32; Albert Abrams, *New Concepts in Diagnosis and Treatment* (1916; rpt. San Francisco: Physico-Clinical Co., 1922), 264–65.

11. Austin C. Lescarboura, "Our Abrams Investigation–V," *Scientific American* 130, no. 2 (February 1924), 87; "The Abrams Cult in Medicine," *Nature* 114, no. 2867 (October 11, 1924), 525–26.

12. Austin C. Lescarboura, "Our Abrams Investigation–VI," *Scientific American* 130 (March 1924), 207–12; Austin C. Lescarboura, "Our Abrams Investigation–VII," *Scientific American* 130 (April 1924), 240, 278–79; "Our Abrams Investigation–XI," *Scientific American* 131 (August 1924), 141.

13. Ralph Lee Smith, "The Incredible Drown Case," *Today's Health* 46, no. 4 (April 1968), 49.

14. *Sun* (Baltimore, MD), January 6, 1951, 1.

15. *Tribune* (Oakland, CA), February 16, 1974.

16. Christopher Tugendhat, *Oil: The Biggest Business* (New York: Putnam's, 1968), 175–76. Richard Webster, in *Dowsing for Beginners* (Woodbury, MN: Llewellyn, 2008), 46, wrote that de la Warr had supposedly found natural gas in the Himalayas of India.

17. O. W. Killam, interview, Oral History of the Texas Oil Industry

Collection, Dolph Briscoe Center for American History, University of
Texas at Austin, May 7, 1956, tape 183.

18. Diana Davids Hinton and Roger M. Olien, *Oil in Texas: The Gusher
    Age, 1895–1945* (Austin: University of Texas Press, 2002), 109–10.

19. O. W. Killam, interview, Oral History of the Texas Oil Industry
    Collection, Dolph Briscoe Center for American History, University of
    Texas at Austin, September 5, 1956, tape 192.

20. Bill Walraven and Marjorie K. Walraven, *Wooden Rigs, Iron Men*
    (Corpus Christi, TX: Corpus Christi Geological Society, 2005), 63.

21. O. W. Killam, interview, Oral History of the Texas Oil Industry
    Collection, Dolph Briscoe Center for American History, University of
    Texas at Austin, May 7, 1956, tape 183.

22. Ibid.

23. Ibid.

24. Bill Walraven and Marjorie K. Walraven, *Wooden Rigs, Iron Men*
    (Corpus Christi, TX: Corpus Christi Geological Society, 2005), 69.

25. "F. Julius Fohs, Discoverer of Mexia Field," *Oil Trade Journal* 12, no.
    11 (November 1921), 80–81.

26. *Evening News* (Mexia, TX), August 13, 1921, 1.

27. *Evening News* (Mexia, TX), August 8, 1921, 2.

28. *Star-Telegram* (Fort Worth, TX), December 25, 1921, 10.

29. *Evening News* (Mexia, TX), January 19, 1922, 1.

30. *News-Tribune* (Waco, TX), December 16, 1921, 7.

31. *Evening News* (Mexia, TX), December 28, 1921, 6.

32. *Star-Telegram* (Fort Worth, TX), November 6, 1921, 36; November
    20, 1921, 37.

33. *Evening News* (Mexia, TX), March 4, 1922, 1.

34. *Standard-Examiner* (Ogden, UT), May 31, 1923, 8; *Tribune* (Oakland,
    CA), March 26, 1922, B-7.

35. *Record-Chronicle* (Denton, TX), July 17, 1977, 19.

36. *Star-Telegram* (Fort Worth, TX), June 4, 1923, 1.

37. *Express* (San Antonio, TX), April 21, 1923, 1.

38. *Star-Telegram* (Fort Worth, TX), June 5, 1923, 1; *Daily News*
    (Galveston, TX), June 6, 1923, 1.

39. *Express* (San Antonio, TX), February 24, 1924, 7.

40. Reno *Nevada State Journal*, March 24, 1939, 9; Roger M. Olien

and Diana Davids Olien, *Easy Money: Oil Promoters and Investors in the Jazz Age* (Chapel Hill: University of North Carolina Press, 1990), 169–70.

41. *Guard* (Eugene, OR), April 19, 1923, 4.

42. *Herald* (Heppner, OR), April 24, 1923, 5.

43. *Morning Register* (Eugene, OR), May 6, 1922, 4.

44. *Daily Capital Journal* (Salem, OR), February 8, 1924, 4; July 23, 1925, 4; July 15, 1927, 4.

45. *Daily Guard* (Eugene, OR), April 20, 1923, 1; April 28, 1923, 1; May 1, 1923, 8.

46. *Morning Register* (Eugene, OR), May 6, 1923, 8.

47. *Mail Tribune* (Medford, OR), January 19, 1924, 3.

48. Richard Sneddon, "The Grapevine," *California Oil World* 56, no. 22 (November 1963), 12; *Morning Register* (Eugene, OR), December 24, 1926, 1; *Mail Tribune* (Medford, OR), January 19, 1924, 3.

49. *Guard* (Eugene, OR), April 21, 1925, 2.

50. *Guard* (Eugene, OR), May 1, 1925, 2.

51. *Guard* (Eugene, OR), December 23, 1925, 12.

52. *Daily Inter-Lake* (Kalispell, MT), April 23, 1928, 4.

53. *Guard* (Eugene, OR), February 9, 1926, 3.

54. *Guard* (Eugene, OR), March 9, 1929, 1.

55. *Guard* (Eugene, OR), July 16, 1928, 1; *Morning Register* (Eugene, OR), July 28, 1928, 7; *Evening Herald* (Klamath Falls, OR), September 7, 1928, 10.

56. *Guard* (Eugene, OR), September 4, 1928, 1; *Morning Register* (Eugene, OR), October 6, 1928, 5.

57. *Morning Register* (Eugene, OR), May 28, 1929, 6.

58. "Geraldine Well Is All Wet – Doodlebug Locates Water Instead of Grease," *Montana Oil Journal*, April 14, 1928 4.

59. *Herald* (Bellingham, WA), May 15, 1931, 3.

60. *Times-Democrat* (Muskogee, OK), October 7, 1913, 4.

61. *Tribune* (Oakland, CA), January 7, 1914, 1.

62. *Tribune* (Oakland, CA), March 19, 1914, 11.

63. Marc I. Pinsel, *150 Years of Service on the Seas*, vol. 1 (Washington, DC: US Dept. of the Navy, 1982), 109–10.

64. *National Cyclopaedia of American Biography*, vol. B (New York: White,

1927), 259.

65. *Sun* (San Bernardino County, CA), July 5, 1933, 2; Don Black, "Moving Lights on Map Trace Planes in the Sky," *Popular Mechanics* 64, no. 3 (September 1935), 346–49.

66. Albert Henry Kingerly, "Strange Radio Devices Locate Buried Treasure," *Popular Science Monthly* 112, no. 4 (April 1928), 158.

67. John C. Cook, "Some Unorthodox Petroleum Exploration Methods," *Geophysics* 24, no. 1 (February 1959), 142–54.

68. Dart Wantland, "The 'Doodlebug' vs. Applied Geophysics," *Mines Magazine* (July 1940), 343.

69. Edgar Cayce Reading 1180-9, Report 1; *Sun* (San Bernardino County, CA), March 15, 1936, 11; *Guard* (Eugene, OR), December 24, 1926, 1; "Spitz Petrolometer Being Demonstrated," *Oil & Gas Journal* 27, no. 20 (October 4, 1928), 148.

70. John C. Cook, "Some Unorthodox Petroleum Exploration Methods," *Geophysics* 24, no. 1 (February 1959), 144–45.

71. *Intelligencer* (Edwardsville, IL), November 4, 1955, 2.

72. *Post-Gazetteer* (Pittsburgh, PA), January 1, 1920, 30.

73. Lincoln *Nebraska State Journal*, October 17, 1927, 1.

74. *Argus-Leader* (Sioux Falls, SD), May 9, 1932, 8.

75. Albuquerque *Morning Journal*, May 22, 1925, 4.

76. John P. Wilson, *Black Gold in the San Simon*, US Bureau of Land Management, Safford, Arizona, District, Report No. 75 (August 1996).

77. *Times Mirror* (Warren, PA), December 28, 1932, 2.

78. *Sentinel* (Woodstock, IL), November 14, 1901, 8; February 9, 1905, 8.

79. *Daily Bee* (Omaha, NE), July 9, 1911, D-1.

80. *Daily Bee* (Omaha, NE), December 2, 1911, 9.

81. *Evening Independent* (Massilon, OH), December 17, 1918, 2.

82. Lincoln *Nebraska State Journal*, November 21, 1926, 5-B.

83. *News* (Frederick, MD), September 4, 1926, 1.

84. Lincoln *Nebraska State Journal*, July 6, 1927, 3; *Journal Star* (Lincoln, NE), August 13, 1926, 1.

85. *Daily Pioneer-Times* (Deadwood, SD), December 23, 1926, 2; March 20, 1927, 1.

86. *Daily Sentinel* (Woodstock, IL), June 24, 1933, 1; Lincoln *Herald*,

May 27, 1932, 3.

87.   Los Angeles *Times*, September 26, 1936, 8.

88.   Los Angeles *Times*, October 5, 1936, section 2, 16.

89.   Reno *Nevada State Journal*, April 8, 1939, 8; *Gazette-Journal* (Reno, NV), December 1, 1938, 11.

90.   *Morning Journal* (Hanford, CA), October 30, 1937, 1.

91.   *Herald* (El Paso, TX), June 23, 1939, 9.

92.   David Hatcher Childress, *The Free Energy Device Handbook* (Stelle, IL: Adventures Unlimited, 1994), 27; Frank Scully, *Behind the Flying Saucers* (New York: Holt, 1950), 185.

93.   "Note on the 'Attractometer,'" *New and Unusual Methods of Petroleum Exploration*, no. 10 (December 1951), 1–3.

94.   John C. Cook, "Some Unorthodox Petroleum Exploration Methods," *Geophysics* 24, no. 1 (February 1959): 144–45.

95.   *Securities and Exchange Commission News Digest*, March 17, 1961, 24–25; *Tribune* (Great Falls, MT), September 27, 1960, 7.

96.   Los Angeles *Times*, June 9, 1964, part III, 15.

97.   Karl Baarslag, *Robbery by Mail* (New York: Farrar & Rinehart, 1938), 233; Norbert R. Mahnken, "No Oklahoman Lost a Penny," *Chronicles of Oklahoma* 71, no. 1 (Spring 1993), 54–55.

98.   M. S. Blackburn, "Radiographic Method of Geophysical Exploration," *World Oil*, August 11, 1947, 43–46; US Patent 1,828,954, October 27, 1931.

99.   *Express* (San Antonio, TX), October 13, 1953, 12A.

100.   *Tribune* (Bismarck, ND), January 2, 1930, 2.

101.   John P. Hickey, ed., *Decisions and Reports*, vol. 2, Securities and Exchange Commission (1939), 744.

102.   John Bluemle, "Prospects and Swindles," *Rocky Mountain Oil Journal*, August 30, 1996, 3, 6.

103.   John P. Hickey, ed., *Decisions and Reports*, vol. 2, Securities and Exchange Commission (1939), 743–47.

104.   *Tribune* (Great Falls, MT), September 27, 1960, 7; *Tribune* (Bismarck, ND), March 27, 1953, 6.

105.   *Transcript* (Freehold, NJ), August 28, 1925, 1; *Gazette* (Billings, MT), November 18, 1921, 12.

106.   Los Angeles *Times*, October 8, 1923.

107. *News* (Miami, FL), August 24, 1925, 16.

108. John P. Hickey, ed., *Decisions and Reports*, vol. 2, Securities and Exchange Commission (1939), 743–47.

109. *News* (Paterson, NJ), August 21, 1925, 13.

110. *Press* (Asbury Park, NJ), October 19, 1927, 3.

111. *Messenger* (Allentown, NJ), February 13, 1930, 4.

112. *Daily Sun* (Brandon, Manitoba), June 7, 1929, 9.

113. *Daily News* (Rhinelander, WI), May 18, 1929, 2.

114. *Daily News* (Rhinelander, WI), October 5, 1929, 2.

115. *Daily News* (Rhinelander, WI), December 12, 1929, 4; January 13, 1930, 4.

116. *Daily News* (Des Moines, IA), November 21, 1904, 2; *Tri-City Star* (Davenport, IA), December 22, 1904, 6.

117. *Register* (Des Moines, IA), December 11, 1938, 7; *Reporter* (Hamburg, IA), January 18, 1934, 5; *Leader* (Chariton, IA), December 16, 1930, 1.

118. *Independent* (Humboldt, IA), October 18, 1923, 1; *Kossuth County Advocate* (Algona, IA), March 29, 1923, 4.

119. Stanley C. Grant, *Summary of Oil & Gas Test Wells in Iowa*, Iowa Geological Survey (November 1976); *Republican* (Humboldt, IA), May 30, 1924, 1.

120. William S. Foster, "Experiments on Rod-divining," *Journal of Applied Psychology* 7, no. 4 (December 1923), 303–11.

121. G. B. Morey, *The Search for Oil and Gas in Minnesota*, Educational Series 6, Minnesota Geological Survey (1984), 24; Minneapolis *Star*, August 14, 1926, 5.

122. *Times* (St. Cloud, MN), April 23, 2005, 19.

123. *Free Press* (Detroit, MI), December 13, 1925, 9.

124. Ida I. Davis, "When the Doodle Bug Balks," *Engineering & Mining Journal-Press* 117, no. 4 (January 26, 1924), 175; *Herald* (Highland Park, CA), January 26, 1923, 2; *Express* (Los Angeles, CA), March 30, 1923, 37.

125. *Times* (Tonopah, NV), November 23, 1923, 1; *Gazette-Journal* (Reno, NV), August 23, 1923, 6.

126. *Cherokee Times* (Gaffney, SC), March 4, 1926, 1; *Ledger* (Gaffney, SC), December 19, 1933, 1; "Local Doings in Goldfield," *Engineering &*

Mining Journal-Press 116, no. 26 (December 29, 1923), 1102; *Free Press* (Detroit, MI), December 13, 1925, 9.

127. *Ledger* (Gaffney, SC), December 19, 1933, 1.

128. *Free Press* (Detroit, MI), December 7, 1938, 4.

129. "Paper read by Lloyd N. Nash August 26, 1937, before the National Association of Better Business Bureaus, Inc.," Securities and Exchange Commission website.

130. "Doodlebug," *Time* 26, no. 17 (October 21, 1935), 60–62.

131. *Herald-Press* (St. Joseph, MI), June 10, 1939, 2; *Free Press* (Detroit, MI), September 7, 1938, 4; June 14, 1939, 7; October 20, 1939, 3.

132. *Journal* (Indianapolis, IN), September 14, 1889, 2.

133. *Pharos-Tribune* (Logansport, IN), July 27, 1897, 18.

134. *Press* (Waterloo, IN), April 13, 1911, 7.

135. "The Divining Rod," *Mining and Scientific Press* 121, no. 21 (November 20, 1920), 721–22.

136. *Liberal* (Lordsburg, NM), July 14, 1921, 1.

137. *Herald* (Montpelier, IN), January 28, 1904, 1.

138. *Daily News-Democrat* (Huntington, IN), June 24, 1903, 1; May 7, 1904, 1.

139. *Herald* (Huntington, IN), July 9, 1909, 1; *Daily News-Democrat* (Huntington, IN), May 31, 1904, 8.

140. *Mail Tribune* (Medford, OR), January 19, 1924, 3.

141. *Daily Record* (Coffeyville, KS), July 2, 1906, 6.

142. *Daily State Journal* (Topeka, KS), September 16, 1911, 8.

143. "Challenge from a Woman," *Oil & Gas Journal*, July 2, 1914, 5.

144. *Daily Tribune* (Chanute, KS), August 7, 1914, 8.

145. *News* (Hutchinson, KS), May 29, 1924, 5.

146. *Star-Telegram* (Fort Worth, TX), September 30, 1967, 6.

147. *Daily News-Record* (Miami, OK), August 18, 1935, 2.

CHAPTER 11

1. George de Mille, *Oil in Canada West* (Calgary, Alberta: Northwest Printing and Lithographic, 1969), 252.

2. *Record* (Agassiz, British Columbia), June 25, 1914, 1; *Oil & Gas Journal*, August 6, 1914, 8.

3. Bakersfield *Californian*, July 24, 1915, 1; *Oil & Gas Journal*, August

5, 1915, 3.

4. *Herald* (Lethbridge, Alberta), August 3, 1923, 9.

5. *Tribune* (Winnipeg, Manitoba), December 29, 1921, 8.

6. "South-eastern Manitoba," *Canadian Institute of Mining and Metallurgy Bulletin* 16, no. 129 (January 1923), 14–15.

7. *Sun* (Vancouver, British Columbia), July 31, 1922, 11.

8. *Sun* (Vancouver, British Columbia), August 18, 1922, 14.

9. *Leader-Post* (Regina, Saskatchewan), May 1, 1926, 30; *News* (Medicine Hat, Alberta), July 30, 1923, 1.

10. *Whig* (Kingston, Ontario), August 24, 1940, 1.

11. *National Post* (Toronto, Ontario), September 6, 1941, 14.

12. *Leader-Post* (Regina, Saskatchewan), August 24, 1985, B2.

13. *Herald* (Calgary, Alberta), August 12, 1959, 25.

14. *Herald* (Calgary, Alberta), November 25, 1946, 17; February 25, 1970, 46; Allen Anderson, *Roughnecks and Wildcatters* (Toronto: MacMillan, 1981), 69–70.

15. *Mining and Engineering Record* 21, no. 5 (September–October 1916), 91.

16. "How Money Is Lost in Wildcat Schemes," *Mining and Engineering Record* 18, no. 3 (October 1912), 34–36.

17. *Sun* (Vancouver, British Columbia), April 13, 1915, 4; June 29, 1915, 5.

18. *Star* (Lincoln, NE), May 14, 1917, 6; *Daily Bee* (Omaha, NE), February 21, 1918, 9.

19. *Daily Sun* (Beatrice, NE), July 16, 1921, 1.

20. *Daily Globe* (Atchison, KS), January 31, 1931, 6.

21. *Age* (Melbourne, Victoria), August 9, 1938, 9.

22. "The Search for Oil," *Mining Review for the Half-Year Ending June 30, 1931*, no. 54, South Australia Department of Mines, 63–67.

23. *Age* (Melbourne, Victoria), July 26, 1938, 9.

24. *Examiner* (Launceston, Tasmania), June 20, 1939, 6; *Age* (Melbourne, Victoria), December 9, 1938, 9.

25. *Lachlander and Condobolin and Western Districts Recorder* (Condobolin, New South Wales), May 30, 1940, 3.

26. *Herald* (Naracoorte, South Australia), January 16, 1940, 4.

27. *Morning Herald* (Sydney, New South Wales), December 3, 1940, 5.

28. "Les Sourciers et le Pétrole Français," *La Nature*, no. 2693 (November 14, 1925), 213.

29. "Scientific Oil Detector," *Oil News* 10, no. 20 (October 20, 1922), 36.

30. Pierre-Louis Maubeuge, *Comme une Odeur de Pétrole* (Pierron, 1996), 40–41, 138; "Aged Frenchman Says He Has System to Locate Oil," *Oil Weekly*, October 21, 1922, 10.

31. *Truant Express and Tumbarumba Post* (Adelong, New South Wales), January 26, 1923, 1.

32. Pierre Fontaine, *L'Aventure du Pétrole Française* (Paris: Fontaine, 1967), 68.

33. "Regards sur l'Avenir de l'Industrie Pétrolière au Maroc," *Revue de Géologie* 15 (1935), 130.

34. *Free Press* (Burlington, VT), January 9, 1930, 19; *Herald* (Lethbridge, Alberta), January 18, 1930, 12; Frederick W. Lee, *Geophysical Abstracts No. 12*, Information Circular 6287, US Bureau of Mines (April 1930), 21.

35. K. Becker, "Vorführungen der Wünschelrute und des Polarisators," *Metall und Erz* 20, no. 6 (March 22, 1923), 101–4.

36. Hans Henning, "Prüfung eines Wünschelrutengängers durch ein Wissenschaftliche Kommission," *Zeitschrift für Psychologie* 32 (1919).

37. Herman A. Holz, "New Process for Locating Mineral Deposits," *Iron Age* (December 9, 1920), 1549.

38. "A German Divining Rod," *Engineering & Mining Journal* 112, no. 27 (December 31, 1921), 1042–43.

39. "The Divining Rod," *Scientific American* 125, no. 13 (September 24, 1921), 214.

40. M. J. M. Bless, "Wichelroede en Geologie: de Schermuly-polarisator," *Miscellanea Geologica Coriovallana*, eds. M. J. M. Bless and J. Van dem Bosch (Aachen, 1981), 13–15.

41. *Report of the Government of the Union of South Africa on South-West Africa for the Year 1924* (1925), 48.

42. Arnold Heim, "Wünschelrute im Schweiz," *Zeitschrift für Angewandte Geologie* 3, no. 8/9 (1957): 426–27.

43. F. W. Lee, "Geophysical Abstracts, No. 1," Information Circular 6120, US Bureau of Mines, (1929), 21–22; A. G. White et al., "Crude Petroleum and Petroleum Products," *Minerals Yearbook 1942*, US

Department of the Interior (1943), 1122; *Diario Carioca* (Rio de Janeiro, Brazil), July 3, 1936, 10.

44. *Petroleum* 25 (1929), 1655.

45. "Divining Rod Tunes in on Ore," *Modern Mechanix* (September 1934).

46. H. Hunkel, "Der 'Radio-Emanator' und Seine Angeblichen Wissenschaftlichen Grundlagen," *Metall un Erz* 30, no. 9, 204–6; *Star* (Windsor, Ontario), December 6, 1924, 19.

47. Otto J. Hansen, "Bedeutende Erfindung für Bergbau und Bohrindustrie," *Allgemeine Österreichische Chemiker und Techniker Zeitung* 45, no. 16 (August 15, 1927), 124; "The Englehardt Mechanical Pendulum," *Petroleum Times* 18 (September 17, 1927), 555.

48. Eric Krlander, *Hitler's Monsters* (New Haven, CT: Yale University, 2017), 271–72.

49. Adolf Schmid, "Device for Detecting Subterranean Waters," United States patent No. 841,188, January 15, 1907.

50. Arthur J. Ellis, *The Divining Rod: A History of Water Witching*, Water Supply Paper 416, US Geological Survey (1917), 24.

51. *Argus* (Melbourne, Victoria), May 29, 1931, 4; J. W. Porter, "Early Surface and Subsurface Investigations of the Western Canada Sedimentary Basin," *Foreland Basins and Fold Belts*, eds. Roger A. Macqueen and Dale A. Leckie, American Association of Petroleum Geologists, Memoir 55 (1992), 125–57; *Herald* (Calgary, Alberta), June 29, 1929, 38; Edgar W. Owen, "Urania, 1925, to Laredo, 1975," *American Association of Petroleum Geologists Bulletin* 60, no. 4 (April 1976), 621–22; J. B. Rathbun, "Gamble in Oil Almost Gone?" *Petroleum Age* 14, no. 7 (October 1, 1924), 25–26.

52. "Vibratory Doodle-bug?" *Oil & Gas Journal* 22, no. 7 (July 12, 1923), 117.

53. "Various Matters," *Petroleum Times* 19 (March 3, 1928), 424; Henri Mager, *Water Diviners and Their Methods*, trans. A. H. Bell (London: Bell, 1931), 55–56.

54. Henri Mager, *Pour Pratiquer la Radio-Physique* (Tours: Deslis, 1938), 71; "Results in the Rhine District," *Petroleum Times* 19 (June 23, 1928), 1166.

55.   Alan D. Chave and Alan G. Jones, eds., *The Magnetotelluric Method* (Cambridge: Cambridge University Press, 2012), 10, 437; *Morning Herald* (Sydney, NSW), February 10, 1931, 11.

CHAPTER 12

1.   James C. Templeton, "Geophysical Prospecting for Oil and Gas," *Petroleum Times* 25 (February 7, 1931), 175; Edgar Wesley Owen, *The Trek of the Oil Finders: A History of Exploration for Petroleum*, Introduction by Wallace Pratt (Tulsa, OK: American Association of Petroleum Geologists, 1975), xiii–xv.

2.   Edgar Wesley Owen, *Trek of the Oil Finders: A History of Exploration for Petroleum* (Tulsa, OK: American Association of Petroleum Geologists, 1975), 516–17.

3.   E. DeGolyer, *The Development of the Art of Prospecting* (Princeton University Press, 1940), 33–36.

4.   Bruce Bryan, "Geophysics—Newest Scientific Aid in the Search for Petroleum," *Oil Bulletin* 16 (February 1930), 200.

5.   J. B. Rathbun, "Gamble in Oil Almost Over," *Petroleum Age* 14, no. 7 (October 1, 1924), 25–26.

6.   *Daily Oklahoman* (Oklahoma City, OK), July 23, 1944, A-9.

7.   Martin Gardner, *Fads & Fallacies in the Name of Science* (New York: New American Library, 1957), 97–98.

8.   *Daily Midget* (Kingfisher, OK), August 15, 1911, 6; oral communication with Kim Gutowsky van der Wal, April 15, 2021.

9.   *Times* (Kingfisher, OK), May 5, 1925, 3.

10.   *Morning Tribune* (Blackwell, OK), October 8, 1926, 12.

11.   Joseph L. Borden, "Developments in Oklahoma during 1941," *Bulletin of the American Association of Petroleum Geologists* 26, no. 6 (June 1942), 1062.

12.   Robert M. Swesnik, "Geology of the West Edmond Oil Field, Oklahoma, Logan, Canadian, and Kingfisher Counties, Oklahoma," *Structure of Typical American Oil Fields*, vol. 3 (Tulsa, OK: American Association of Petroleum Geologists, 1948), 361–64.

13.   Claude V. Barrow, "Main Street . . . and Oil," *Daily Ardmoreite* (Ardmore, OK), November 11, 1962, 7A.

14.   *Evening News* (Ada, OK), April 13, 1943, 3.

15. *Interstate Oil Compact Quarterly Bulletin* (October 1944), 12–43; *Oil & Gas Journal*, March 25, 1943, 346; April 15, 1943, 155; April 22, 1943, 110; July 15, 1943, 109; February 17, 1944, 107; April 27, 1944, 135; May 11, 1944, 159; May 18, 1944, 113.

16. *Telegram* (Salt Lake City, UT), March 25, 1949, 32; Oklahoma City *Daily Oklahoman*, September 8, 1943, 12.

17. Jim Dent, *Monster of the Midway* (New York: St. Martin's, 2003), 198–206.

18. *Valley Morning Star* (Harlingen, TX), May 5, 1946, 1; April 6, 1951, 5.

19. A. Newton Plummer, *The Great American Swindle* (New York: A. N. Plummer, 1932), 269–82; Henry P. Stokes, et al., *History of the Class of 1909, Yale College* (New Haven, CT: Yale University Press, 1915), 193–94.

20. *Time*, November 10, 1930, 26.

21. *New York Times*, April 8, 1932, 39; June 8, 1932, 28; March 24, 1933, C26; April 12, 1935; May 1, 1935, 29.

22. *New York Times*, July 9, 1931, 6; *Herald* (Syracuse, NY), September 2, 1934, 3; October 28, 1934, section 2, 1; August 10, 1938, 3; New York *Daily News*, July 9, 1931, 2.

23. *Times* (Reading, PA), February 26, 1935, 4.

24. *New York Times*, March 1, 1937, 19.

25. *Oil & Gas Journal*, March 9, 1933, 52; June 22, 1933, 119.

26. *Oil & Gas Journal*, August 5, 1943, 68; October 7, 1943, 108.

27. *Republican* (Council Grove, KS), June 2, 1947, 1; June 25, 1947, 1; Santa Fe Pacific Railroad Co., Well files of the New Mexico Oil Conservation Commission, 30-031-05404, 30-031-05410.

28. *Post* (Denver, CO), October 15, 1952, 29; October 16, 1952; *Independent* (Gallup, NM), July 18, 1951, 2.

29. *1947–48 Rocky Mountain Petroleum Yearbook* (Denver: Petroleum Publishers, 1948), 21–23.

30. *Post* (Denver, CO), October 20, 1952, 1. San Francisco *Chronicle*, January 20, 1954, 19; *Arizona Republic*, March 23, 1954, 8; June 13, 1954, 3. *Daily Tribune* (Greeley, CO), April 13, 1954, 1.

31. FBI memoranda, November 18, 1952, 4; February 2, 1953, Silas Newton file, FBI FOIA website.

32.   *Rocky Mountain News* (Denver, CO), November 18, 1953, 21.

33.   Bakersfield *Californian*, February 14, 1949, 20.

34.   Quoted in Chris B. Evans, *Alien Conspiracy* (Ceredo, WV: Alarm Clock, 2003), 97.

35.   Frank Scully, *Behind the Flying Saucers* (New York: Holt, 1950), 32–33, 127.

36.   Karl T. Pflock, "What's Really Behind the Flying Saucers? A New Twist on Aztec," *The Anomalist*, no. 8 (2000), 137–61.

37.   Karl T. Pflock, *Roswell* (Amherst, NY: Prometheus, 2001), 85–93.

38.   Letter dated December 1, 1949, from Lt. Col. Keefe (USAF) to Commanding General, Wright-Patterson Air Force Base, Karl Pflock UFO Collection, Ohio State University, Columbus; Frank Scully, "Scully's Scrapbook," *Variety*, October 12, 1949, 61; November 23, 1949, 24.

39.   Frederick K. Ungerer, "Unconventional Aircraft (Flying Saucers from Venus Come to Earth)," US Air Force, Report of Investigations, January 23, 1950, Karl Pflock Collection, Ohio State University, Columbus.

40.   *Post* (Denver, CO), March 12, 1950, 3A; March 17, 1950, 1.

41.   Frank Scully, *In Armour Bright* (New York: Chilton, 1963), 197, 204.

42.   Most of p. 155 and the top of p. 156 of Scully's book are plagiarized from p. 103 of J. J. Jakosky, *Exploration Geophysics* (Los Angeles: Trija, 1950).

43.   *Post* (Washington, DC), September 10, 1950, B5.

44.   Roland Gelatt, "In a Saucer from Venus," *Saturday Review*, September 23, 1950, 20–36; "Saucers Flying Upward," *Time*, September 25, 1950, 75–76.

45.   *True* (September 1952), 17–112; letter from Frank Scully to Paul Smith, June 19, 1951, Scully papers, American Heritage Center, University of Wyoming, Laramie, box 3, folder 1; J. P. Cahn, sound recording of a public lecture, February 14, 1984, Campbell, CA

46.   *Chronicle* (San Francisco, CA), January 17, 1954, 1.

47.   *Avalanche* (Lubbock, TX), November 14, 1950, 8; Bob Considine, "The Disgraceful Flying Saucer Hoax," *Cosmopolitan* (January 1951), 32–102.

48.   Quoted by J. P. Cahn in "Flying Saucer Swindlers," *True* (August

1956), 69–72.

49.   *Post* (Denver, CO), October 20, 1952, 1.

50.   *Rocky Mountain News* (Denver, CO), October 19, 1952, 5.

51.   *Daily Times* (Greeley, CO), February 12, 1953, 7.

52.   *Rocky Mountain News* (Denver, CO), November 14, 1953, 6.

53.   *Post* (Denver, CO), December 17, 1953, 3; December 22, 1953, 3.

54.   *Rocky Mountain News* (Denver, CO), December 18, 1953, 22.

55.   *Rocky Mountain News* (Denver, CO), December 22, 1953, 10; December 9, 1953, 30; December 23, 1953, 50.

56.   *Post* (Denver, CO), June 14, 1954, 3; *Rocky Mountain News* (Denver, CO), June 15, 1954, 10; *Star* (Tucson, AZ), June 15, 1954, 2; "Order Granting Probation," *People of the State of Colorado v. Leo A. GeBauer and Silas M. Newton*, case #40891, Colorado 2nd Judicial District, Denver, Colorado (n.d.).

57.   *Rocky Mountain News* (Denver, CO), December 30, 1953, 19; June 15, 1954, 10.

58.   *Post* (Denver, CO), April 26, 1959, 24A; letter from Silas Newton to Harold Sherman, March 9, 1971, Harold Sherman Collection, box 9, file 4.

59.   *Rocky Mountain News* (Denver, CO), December 15, 1956, 31.

60.   Ibid., February 8, 1955, 42; March 12, 1958, 8; October 1, 1959, 8; *Post* (Denver, CO), March 12, 1958, 1.

61.   *Daily Citizen* (Tucson, AZ), September 3, 1957, 4.

62.   Letter from Silas Newton to Harold Sherman, March 9, 1971, Harold Sherman Collection, University of Central Arkansas, Conway, Collection M87-8, Series 1, Subseries 6, box 9, file 4.

63.   *Rocky Mountain News* (Denver, CO), October 1, 1959, 8; *Daily Sun* (Yuma, AZ), August 29, 1962, 5.

64.   *Arizona Daily Sun* (Flagstaff, AZ), September 2, 1959, 1; September 10, 1959, 1; Patricia Hartwell, "Oilman Confident Sedona Limestone Offers Big Yield," *The Arizonian*, July 17, 1964, 3.

65.   *Arizona Republic* (Phoenix, AZ), June 25, 1966, 26; well file for the Harless #1 Federal oil well, API #02-025-00169, Arizona Geological Survey website.

66.   Letter from Silas Newton to Harold Sherman, September 7, 1968, Harold Sherman Collection, box 9, folder 3.

67. Letter from Harold Sherman to Ludwig Schraut, January 6, 1969, Harold Sherman Collection, box 7, folder 17.

68. "Assay of Ore," Eisenhauer Laboratories, Los Angeles, to S. M. Newton, Harold Sherman Collection, box 9, file 2; FBI memoranda, October 22, 1968, September 30, 1970, FBI FOIA website.

69. Letter from Silas Newton to Harold Sherman, September 7, 1968, Harold Sherman Collection, box 9, file 3.

70. Memo dated November 30, 1970, FBI FOIA website.

71. Letter from Harold Childers to Al Pollard, June 27, 1974, Harold Sherman Collection, box 9, file 5.

72. William L. Moore, *Crashed UFOs: Evidence in the Search for Proof* (Burbank, CA: self-published, 1987), 7.

73. Karl T. Pflock, "What's Really Behind the Flying Saucers? A New Twist on Aztec," *The Anomalist*, no. 8 (2000), 137–61; *Daily Times* (Farmington, NM), March 20, 2002, 1.

74. This chapter relies heavily on the work of Pierre Péan, the journalist who broke the story in *Le Canard Enchaîné* and later in his book *V: Enquête sur l'Affaire des "Avions Renifleurs"* ["V: Report on the Affair of the 'Sniffing Planes'"] (Fayard, 1984), 21–25, 31, 223.

75. *Post* (Birmingham, England), September 7, 1961, 8; *Independent* (Pasadena, CA), August 25, 1961, 10.

76. Daniel Singer, "Auto Workers and 'Sniffing Planes,'" *Nation* 238, no. 7 (February 25, 1984), 218–21.

77. "Les Photos de l'Illusion," *Paris Match*, January 27, 1984, 28.

78. *Philadelphia Inquirer*, February 9, 1984, F-1.

79. Pierre Péan, *V: Enquête sur l'Affaire des "Avions Renifleurs"* (Fayard, 1984), 124–27.

80. Robert Hutchison, *Their Kingdom Come* (New York: St. Martin's, 1997), 160–61.

81. Jorge Navarro Comet, "The Great Oil Sniffer Hoax," *AAPG Explorer* 39, no. 8 (August 2018), 20.

82. Christophe Carrière and Jean-Marie Pontaut, "Aldo Bonassoli: Le Roi du Pétrole," *L'Express International*, no. 3236 (July 10, 2013), 52–55.

83. Pierre Péan, *V: Enquête sur l'Affaire des "Avions Renifleurs"* (Fayard, 1984), 172.

84. "The Great Oil Sniffer Hoax," *Discover* 5, no. 3 (March 1984), 8.

85. The incident is reminiscent of R. W. Wood's 1904 trickery to prove that N-rays were a product of French physicist Prosper-René Blondlot's imagination: *Nature*, September 29, 1904, 530–31.

86. David Dickson, "Oil Prospectors Make a Strike in Paris," *Science* 223, no. 4633 (January 20, 1984), 261–62.

87. François Gicquel, "Rapport Confidential de la Cour des Comptes Sur Certaines Operations de l'ERAP," January 21, 1981; Marci McDonald, "Sniffing Out a Coverup," *Maclean's* 97 (February 6, 1984), 23.

88. "Big Stink," *Time*, January 30, 1984, 27.

CHAPTER 13

1. "Number of Wells: US Geological Survey," *Mineral Resources of the United States* (1908–1931); US Bureau of Mines *Minerals Yearbooks* (1932–1948); US Energy Information Administration website (1949–2010).

2. Anne Jaeger-Nosal, *Les Chercheurs d'Eau* (Geneva, Georg, 1999), 20.

3. Tim Archer and Greg Hodges, "New Geophysics? Or Bad Geophysics? How Not to Be Fooled," Technology Assessment Group, http://www.reid-geophys.co.uk/wp-content/uploads/2017/10/Technology-Assessment-Group-Flyer.pdf.

4. Richard P. Feynman, "Cargo Cult Science," *Engineering and Science* (June 1974), 10–13.

5. Matthew J. Sharps, "Percival Lowell and the Canals of Mars," *Skeptical Inquirer* 42, no. 3 (May–June 2018), 41–46.

6. Willy Ley, *Mariner IV to Mars* (New York: New American Library, 1966).

7. Irving Langmuir, "Pathological Science," *Physics Today* (October 1989), 36–48.

8. C. O. Hage, "Memorial, George Sherwood Hume," *Bulletin of Canadian Petroleum Geology* 14, no. 1 (March 1966): 191–92.

9. George S. Hume, "Confidential," four-page memo, August 29, 1960, Glenbow Collection, University of Calgary Library archives, Ned Gilbert fonds, M-8743, box 28, file 285.

10. Ned Gilbert, "Story of a Black Box," *Petroleum History Society Archives* 16, no. 1 (February 2005), 6–8.

11. Susan Birley, interview with Charlie Dunkley, May 14, 1984, transcript, Petroleum Industry Oral History Project, Glenbow Collection, University of Calgary Library archives.

REFERENCES

Bird, Christopher. *The Divining Hand*. New York: Dutton, 1979.

Boatwright, Mody C. *Folklore of the Oil Industry*. Dallas: Southern Methodist University Press, 1963.

Clark, James A., and Michel T. Halbouty. *The Last Boom*. New York: Random House, 1972.

DeGolyer, E. *The Development of the Art of Prospecting*. Princeton University Press, 1940.

Ellis, Arthur J. *The Divining Rod: A History of Water Witching*. US Geological Survey, Water Supply Paper 416, 1917.

Olien, Roger M., and Diana Davids Olien. *Easy Money: Oil Promoters and Investors in the Jazz Age*. Chapel Hill: University of North Carolina Press, 1990.

Péan, Pierre. *V: Enquête sur l'Affaire des «Avions Renifleurs»* (V: Report on the Affair of the "Sniffing Planes"). Paris: Fayard, 1984.

Penrose, Evelyn. *Adventure Unlimited*. London: Spearman, 1958.

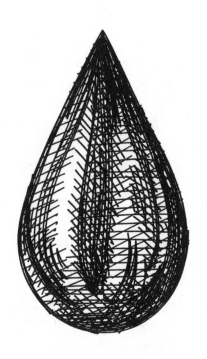

INDEX

Abner Davis Millionaires' Club, 165
Abner Davis Trustee Plan, 164
Abrams, Albert, 138–39, 243
affinity instrument, 145, 158–59
AGIP, 113
Alaska, cities in: Nome, 86; Point Barrow, 180
Albert, Eddie, 54, 57–58
Alberta, Canada, cities in: Baron, 110; Calgary, 3, 109, 138, 181, 184, 242; Drumheller, 185; Edmonton, 110
Alberta, Canada, oil fields in: Baron, 110; Drumheller, 185; East Calgary, 243; Mitsue, 244; Red Earth, 244; West Drumheller, 185
Alexander, E., 72
Amerada Petroleum, 89
American Dowser, 96
American Society of Dowsers, 64
American Technical University, 156
Amoco (Stanolind), 210, 212
Anderson, Robert C., 55–59
Anglo-American Oil, 118
Ankarlo, George, 158
Annemann, Theodore, 73
anticline, 82–83
Arizona, cities in: Casa Grande, 180; Flagstaff, 242; Phoenix, 53–54, 180, 213–14, 222; Sedona, 223–24; Tucson, 215; Wickenburg, 180
Arizona Oil and Gas Commission, 224
Arizona State University, 54
Arkansas Valley Oil Company, 76
Artesian Well Company, 14
Arthur (senator), 188

Ashmore, G. Percy, 118
Associated Press, 211
Association for Research and Enlightenment (A.R.E.), 41–42
astrology, 63
Atkins, J. E., 87
Atlantic University, 42
Atlantis (lost continent), 27
Attractometer, 161–64
Austria, cities in: Schleinbach, 116; Vienna, 117

Bacon, Francis, 91
Baker, Benjamin, 209
Balcones Fault, 131
Barbiou, Narcisso, 120
Barnsdall, William, 17–18
Barrabé, Louis, 111
Barre, Raymond, 235
Battle Creek Enquirer, 50
Bayne, Samuel Gamble, 68
Beamer, Paul, 211
Behind the Flying Saucers, 217–18
Bein, Thomas, 223
Bell, Arthur, 118
Bennett, M. V., 183
Benson, Amanda, 18
Bentley, George, 179–80
Berger, Jacob, 171
Bertschy, Adolph J. P., 159–61
Biddleford, Maine, 91
Big Bonanza, The, 18
billet reading, 72
biogeophysical method, 121
biolocation, 120–21

biophysical method, 121
Bird, Christopher, 98–100, 121
black box, 3–4, 136, 139–40
Black River Limestone, 51
Blackburn, McMaster Sylvester, 145, 164–65
Blake, Amanda, 54
Blau, Ludwig, 138
Blazer Corporation, 103
Blunt County, Alabama, 41
Boatright, Mody, 132
Bockhold, John, 209–10, 212
Boise, Idaho, 146
Bonassoli, Aldo, 227–36
Booch Sand, 79
Boogher, Chet, 65, 72
Booher, John S., 69–72
Booram, Isaac, 18
Boothby, F. L. M., 118
Boulder County, Colorado, 78
Boulder Oil Company, 78
Bourland, B. D., 204
Bowman, Levi Mack, 161–64, 166
Boyer, Daniel, 233, 236
Brewer, Lizzie, 18
Brewer, Watson and Company, 12
British Columbia, Canada, cities in: Pitt Meadows, 185; Vancouver, 181, 183, 187; Victoria, 185
British Society of Dowsers, 118
Broman, Francis, 217
Brown, Paul Clement, 98–100
Brown, Thomas H., 67
Brunton compass, 83
Brussels, Belgium, 227
Buchanan, Arthur, 25
Buffalo Morning Express, 17
Buffalo Oil and Gas Company, 186
Bulletin of Canadian Petroleum Geology, 242
Bulletin of the American Association of Petroleum Geologists, 204
Burbank, William B., 174
Burmah Oil Company, 140
Burt, R. E., 127, 130–31

Burton, E. O., 87
Buzzascope, 157
Byrdstown, Tennessee, 101

Cabot, Bruce, 217
Cabot, Laurie, 53
Cahn, John P., 218–20, 223–24
Cal Tech University, 213
Calgary Albertan, 181
Calhoun, Dorcie, 8–10
California, cities in: Darwin, 93; Encinitas, 157; Hollywood, 57, 162, 206, 215; Johannesburg, 86; Long Beach, 176; Los Angeles, 19, 55, 99, 155, 161–62, 166–67, 173, 180, 215, 219, 223, 225–26, 242; Oakland, 117, 132, 155–56, 178; Pasadena, 19, 173; Rancho Cucamonga, 173; Randsburg, 93; Red Bluff, 178; Richmond, 178; San Bernardino, 100; San Francisco, 49, 147, 218; San Juan Capistrano, 157; Santa Ana, 179
California, counties in: Los Angeles, 98; Napa, 178; San Bernardino, 157; San Diego, 157
California, oil fields in: Dominguez, 99; Signal Hill, 119, 173
Campbell, A. T., 205–7
Campbell, Lucretia, 178–79
Canadian Pacific Oil Company, 181
Canard enchaîné, Le, 235, 284n74
Canfield, George B., 177
Canfield, Ira, 75–77
Canfield, Isaac, 74–79
Canfield Syndicate, 79
Cardiff, Wales, 118
Carpenter, Everett, 83
Carrillo, Ramon, 179–80
Carter Oil, 84
Cassiday, Alexander, 76
Cayce, Edgar, 27–43, 93, 157
Cayce, Gertrude, 40
Cayce Energy Company, 42
Cayce Operating Company, 42
Cayce Petroleum Corp., Inc., 42

Cazenovia Republican, 15
Central Intelligence Agency (CIA), 54
Chadwick, Arbie R. ("Mighty Red"), 48
Chambers, Howard, 64
Chanute Tribune, 179
Chase Oil, 202
Chattanooga, Tennessee, 55–59
Chemigraph, 144
Chevron (the California Company), 99, 210, 244
Chi Oil, 202–3
Chicago Tribune, 71
Choat, John, 208
Cities Service, 83, 242
Citizens Gas and Oil Mining Company, 176
Clark, Dutch, 205
Cleburne Morning Review, 28
Clemens, Samuel, 18
Clementsville Oil and Gas, 101
Cleo, Merselle, 144
Cloakey, George, 242–44
Cofer, C. L., 151, 178
Colgrove, Chester W., 162
Collier's, 9
Colorado, cities in: Boulder, 75, 77; Canfield, 75; Cañon City, 75–76; Colorado Springs, 77, 132; Cripple Creek, 85; De Beque, 79; Denver, 125, 157, 159, 207, 209, 212, 214, 217–20, 223; Florence, 75, 77; Fort Collins, 78; Greeley, 75; Pueblo, 85; Telluride, 60; Trinidad, 100
Colorado, oil fields in: Boulder, 78, 80; Cañon City, 76–77; Florence, 76–77, 80; Rangely, 209–13, 215, 223
Colorado College, 205
Colorado Group, 76
Colorado School of Mines, 157, 200, 221
Columbia University, 73
Comer (storekeeper), 70
Conser, Jerry, 42, 54
Cook, Frederick, 146
Cook, G. W., 144
Coquette well, 7–8

Corsun, Herman, 220
Cottonwood Oil Company, 224
Cox, John F., 178
Cox, Seymour E. J., 146
creekology, 46
Crocker, Frederick, 68
Crosley, Daniel, 13
Crosley, James, 13
Crown Oil Company, 153
Cunningham-Craig, E. H., 118
CUPET, 120
Curie, Pierre, 184

D Sandstone, 95
D'Aigneaux, Georges Paul, 184
Dade County, Florida, 36
Dahlgren, John, 90–91
Daisy Bradford #3 well, 44–46
Daum, Arnold R., 81
Davis, Abner, 145, 164–65
Davis, Chet, 99
Davis, Edgar B., 31–32
Davis, Lee, 98
Davison, W. C., 171
Day, Doris, 57–58
Day, Edith, 10
De la Warr, George, 139–40
De Murist, Florian Bourqui, 112
De Villegas, Alain, 227–36
De Weck, Philippe, 229, 231
Deep Rock Oil, 46
Delkeskamp, Rudolf, 113
Denver Post, 217
Derricks of Destiny, 68
divining. *See* dowsing
Divining Hand, The, 100
Dixie Oil, 101
Dixon, Jeane, 48–49, 60
Doheny, Edward, 81
Dome Petroleum, 243–44
doodlebug, 4; decline in use, 207, 238–41; definition, 124–25; first use of term, 132–34; increase in use, 136–38, 144
dowsing, 13; bobber, 64; decline in

use, 237–38; definition, 63; L-rods, 64–65; map dowsing, 64, 114, 120–21; opposition to, 65, 84–85, 113, 116; pendulums, 64–65, 85–87, 98–100, 104, 106, 112; physical sensation, 64–65, 70–72, 110, 113, 119; Y-rods, 64–65, 74–80, 103–4, 106, 110

Drake, Edwin, 11, 67, 109

Drake well, 12, 17

Dreamfield Oil Company, Ltd., 10

dreams, 7–10

Driver, William, 168–69

Dukakis, Michael, 53

East Texas Oil Museum, 44

École Polytechnique, 233, 235

Edgar   Cayce   Petroleum   Company, 30–31, 33

Edinburgh, Scotland, 118

Edison, Thomas, 136–37, 159

Edwards (doodlebugger), 166

Egbert, Les, 95–96

Einstein, Albert, 221

Eisenhower, Dwight, 90

El Dorado, Arkansas, 43–44, 160

Electro-Magnetic Mineral Indicator, 176

Electrostatic Balance, 160–61

Elf, 227, 229–36

Elliott (businessman), 85

Ellis, Arthur, 198

Empire Gas and Fuel, 83

Engineering and Mining Journal, 90, 92

England, cities in: Brighton, 117; Oxford, 140

ERAP, 235

Eremeev, A. N., 121

Estinès, Pierre, 189

Ethereal   Wave   Petroleum   Magnet, 183–84

Ethridge, John H., 109

Eugene Divinity School, 147–51

extrasensory perception (ESP), 47, 59, 73

Falkinger, Hans, 117

false memory, 73

Favier (dowser), 120

Fawcett Publications, 223

Federal Bureau of Investigation (FBI), 55, 162, 213–14, 217, 220

Fenley, Guy, 23–24

Fenley, Maggie Wilson, 24

Feynman, Richard, 241, 244

Finkenbiner, John S., 177–78

Fish Lake Merger Oil Company, 173

Flader, Herman, 214, 220–23

Flint Rock Oil, 202

Florida, cities in: Cassadaga, 60; Cocoa Beach, 104

Floridan aquifer, 36

Fohs, F. Julius, 26, 143

Fontaine, Pierre, 190

Ford, Henry, 137

Fort Collins Express, 78

Fortin (Abbé), 194

Foyo (dowser), 120

France, cities in: Paris, 112, 184; Puy-de-Dôme, 189

France, oil and gas fields in: Bonrepos oil field, 230; Castera Lou oil field, 230, 232; Gabian oil field, 111; Lacq gas field, 234; Lannemezan oil field, 230; Parentis oil field, 190; Soudron oil field, 233

Frank (doodlebugger), 185

Franklin, Frania Tye, 44

Franklin Institute, 156

Friedl, Karl, 117

Friesch, Mable, 211–12

Friesch, Wenzel, 211–12

Fulton, Benjamin F., 174–76

Gagnebin, (Prof.) Elie, 111

Galileo, 149

Gandy, J. S., 159

Gardner, Alice B., 19

gas witch, 107

GeBauer, Leo, 207, 212–16, 219–22

Gee, Worthy, 165

Gehman, Beatrice Anne, 60

General Lee Oil Company, 145–46
Geologia Rudnykh Mestorozhdenii, 121
Geological Survey of Canada, 4, 242
Geological Survey of the Netherlands, 193
geologists, 5–6, 69, 81–82, 99, 103, 114
    employment of, 81–82
    opposition to, 82–84
Geophysical Survey Syndicate, 163
geophysicists, 47, 99, 114, 199–200
Geophysics, 158
Geoscope, 194
Geraldine Oil Company, 154
German Academy of Science, 114
German National Association for Dowsers, 114
Germany, cities in: Frankfurt, 192; Hamburg, 114
Germany, oil fields in: Reitbrook oil field, 114; Wietze oil field, 110, 114
Getty, J, Paul, 131
Getty Oil, 99
Getz, John, 171
Giscard d'Estaing, Valéry, 233–36
Gilbert, Ned, 3–4, 244
Giraud, André, 234
Globe (Atchison, Kansas), 187
Globe Petroleum Trust, 87
Goodyear, Charles, 137
Göring, Hermann, 194
Gould, Charles N., 135
Grange, Red, 205
Grayburg Oil, 208–9
Great Eastern Oil, 188
Great Plains Development Company, 185
Great Plains Oil, 184
Green Acres, 54
Griffith, (Dr.) Peyton Standifer, 126–32
Griggs, Joe, 18
Grober, George, 39
Gross, Henry, 91–96
Guaranty Oil Company of Oregon, 148–53
Guerenneur, Pierre, 190
Guest, William M., 172–74

Guillaumat, Pierre, 229
Gulf Oil, 83, 204
Gunsmoke, 54
Gutowsky, Assaph ("Ace"), 201–6
Gutowsky, Augusta Ladwig, 202
Gutowsky, Leroy ("Ace"), 205–6
Gutowsky van der Wal, Kim, 203
Gypsy Oil Company, 83, 204

Halas, George, 205
Hall, T. D., 144
Hamilton County, Indiana, 106
Hammond, John Hays, 192
Hanson, Herman, 165–66
Happich, L., 114
Harding, W. C., 78
Harless, Richard, 223–24
Harmonial wells, 15–16
Harrisburg News, 86
Hartford City Telegram, 106
Harvard University, 73
Hawkins, W. Taylor, 106–7
Hayden, Ferdinand, 75, 77–78
Hayes, Reuben C., 110
Heim, Arnold, 112, 193
Henry Gross and His Dowsing Rod, 91–92
Herman Hanson Syndicate, 163, 165–67
Herzog, Emerich, 117
Heseman, L. F., 65
Hibbing-Sunburst Oil Company, 154
Hickey, Frank, 222
Hieronymus machine, 138
Hill Oil, 83
Hillsdale County, Michigan, 50–51
Himmler, Heinrich, 194
Hirschi, Hans, 81
Hobbs, Aaron, 25
Hollett, Clarence Elihue, 103–4
Holtz, Herman, 192
Horowitz, Jules, 234
Houdini, Harry, 73
Houseknecht, Annette, 50–51
Houseknecht, Blanche, 50
Houseknecht, Ferne, 50–52
Houseknecht, George (younger), 50–52

Houseknecht, George (elder), 50
Houseknecht, Helen, 50
Houseknecht, Luluah, 50
Houseknecht Oil, 52
How to Make a Million Dowsing and Drilling for Oil, 102
Hughes, Howard, Sr., 25
Hughes, J. K., 25
Humble Oil Company, 84, 131, 138
Hume, George S., 4, 242–44
Humphreys, (Col.) A. E., 25–26, 142–44
Hunt, Haroldson Lafayette, Jr., 43–49
Hunt, Haroldson Lafayette, III, 48–49
Hunt, Lyda Bunker, 44
Hunt, Margaret, 47
Hunt, Nelson Bunker, 49
Hunt, Ray Lee, 49
Hunt, William Herbert, 47, 49
Hunt Oil, 49
Hunter, A. M., 78
Hunton Limestone, 204
Hurley, Wilson, 104
Hutchinson News, 87

Illinois, cities in: Chicago, 14–15, 17, 177, 202, 204, 218; Jacksonville, 21; Mattoon, 177; Zion, 170
Illyes, Jacob R., 106
Imperial Oil Company of Kansas, 94
In Armour Bright, 218
Independent Oil and Financial Reporter, 167
Index (Frankfort, KS), 187
Indian Mound Oil Company, 186
Indiana, cities in: Anderson, 71–72; Cambridge City, 71; Fort Wayne, 71; Hagerstown, 71; Huntington, 71; Muncie, 177; Parker City, 177; Portland, 174, 176; Plymouth, 103–4; Valparaiso, 21, 23
Indiana Oil and Gas, 208–9
Indiana Southwestern, 208
Instrument Exploratory Association, 207
International Academy of Psychobiophysics, 227

International Royalty Holding Company of Canada, 154
Iowa: cities in: Central City, 53; Clarinda, 170–71; Council Bluffs, 160; Dubuque, 53; Humboldt, 171; Manilla, 170
Iowa, counties in: Linn, 53; Montgomery, 103
Ireland, Richard, 53–55
Iron Age, 192
Iverson, Clarence, 89–90
Ives, Burl, 57

J Sandstone, 95
Jackson, George W., 25
Jackson County Oil and Gas Company, 133
Jacobi, Belle/Alice, 169–70
Jaeger-Nosal, Anne, 64, 110
James, Abraham, 14–17, 19
Jamieson, Thomas J., 10
Jeannet, Jean, 190
Jewell, John W., 101
Johns Hopkins Applied Physics Laboratory, 92
Johnson, Rudolph, 42, 93
Joiner, Columbus Marion, 44–46

Kahn, David, 28, 30, 249n35
Kalispell-Niarada Oil Company, 154
Kansas, cities in: Arkansas City, 106; Beattie, 187; Chanute, 178; Hutchinson, 179–80; Hoisington, 159; Oak Mills, 187; Russell, 160; Salina, 102; Wichita, 23, 87, 103, 133, 187, 189; Winchester, 186–87
Kansas, counties in: Anderson, 180; Montgomery, 179; Morris, 110; Nemaha, 187; Rice, 187; Russell, 87
Kansas, oil fields in: Augusta, 83; Eldorado, 83; Isern, 189; Raymond, 189; Shurr, 187, 189; Welch-Bornholdt, 180
Keeler, William Wayne, 59–61
Keeler Investments, 61

Kennebunkport, Maine, 92
Kennedy, John F., 55
Kentucky, cities in: Bowling Green, 3; Clementsville, 101; Scottsville, 101
Kentucky, counties in: Allen, 13; Breathitt, 95; Lincoln, 101; Warren, 41
Kentucky Derby, 48–49
Kepler, Aaron C., 7–8
Kepler, George M., 7–8
Ketchner, Casper, 10
Kevin-Sunburst oil field, 109, 154, 183
Keyes, Tom, 184
Killam, Oliver Winfield, 140–42
Kimball, LaVergne J., 175–76
Kingfisher Daily Midget, 202
Kingfisher Farmers' Oil Company, 202
Kleyhauer, Alfred, 213, 220, 223
Knowles, Ruth, 47
Koehler, Friedrich E., 130
Koehler, George, 217
Konnov, Yu., 121–22
Kozyanin, V. K., 121
Kuebelbeck, Jim, 64
Kump, A., 114

Ladwig, Theodore, 202
Langmuir, Irving, 241, 244
Larkin, Zulah, 50–52
Laster, Ed, 44
Le Honda, C., 144
Lee, Robert Aaron, 145–47
Lee, Robert E., 145
Leidy Prospecting Company, 9
Lemuria (lost continent), 27
Lenard, A. G., 88
Lenora Mining and Milling Company, 91
Lenz, Leonard, 172
Leonard, Arthur, 165–66
Lethbridge Herald, 10
Lewis, E. G., 84
Lily (spirit medium), 144
Limerick, Saskatchewan, Canada, 184
Lind, Boyd V., 159
Lindbergh, Charles, 209

Lions Club, 56
Live Oak Oil Association, 32–33
Lloyd, A. D. ("Doc"), 44–46
Lockett, James, 2
Logansport Pharos-Tribune, 175
Long, Jacob, 105–6
Long, Joseph, 28
Louisiana, cities in: Mansfield, 158; Shreveport, 44, 163
Louisiana, parishes in: Acadia, 47; Morehouse, 41
Louisiana, oil fields in: Maxie, 47; Ship Shoal Block 207, 49
Lowell, Percival, 241–42
Lubeck Oil Company, 104
Lucas, Anthony F., 129
Luke Wilson Oil Company, 86–87
Luogo, Augusta Del Pio, 113
Lurano, Italy, 227

Macksburg oil field, 69
Mager, Henri, 111, 197–98
Magnetic Oil Company, 86
Magnum, David, 42
Manchester Guardian, 117
Mancos Shale, 209
Mangum Star, 105
Manitoba, Canada, cities in: Stonewall, 187; Winnipeg, 183
Mansfield Automatic Oil Finder, 186, 194–98, 221
Mansfield News Journal, 97
Mardonov, V. N., 122
Marrion, W. E., 153
Maryland, cities in: Baltimore, 201; Laurel, 92
Marylees, K. W., 118
Massachusetts, cities in: Boston, 31; Salem, 53, 92
Massachusetts Institute of Technology (MIT), 98
Matacia, Louis, 64
Mathhieu (dowser), 63
Matveev, V. S., 121
Maubeuge, Pierre-Louis, 190

McCleary, Wilbur, 132–34
McClintock, Hamilton, 11
McMan Oil, 83
Medical World, 17
Melbourne, Australia, 188
Merideth, George T., 221
Mermet, Alexis, 64, 111–13, 144
Meseck, Wilhelm, 114
Messing, Wolf, 120
Mexico City, D. F., 58
Miami Oil and Natural Gas Company, 36
Miami Sentinel, 60
Michigan, cities in: Coldwater, 50; Detroit, 31, 50, 78, 172–74, 205
Michigan, oil fields in: Albion-Scipio, 50–52; North Adams, 52; Saginaw, 160
Midlothian Petroleum Syndicate, 118
Millennium Minerals, 55
Miller, Wesley C., 4, 242–43
Millikan, Robert, 213
Mineometer, 185
Mineral indicator, 166–69
Mining and Engineering Record, 185
Minkler, Robert, 99
Minnesota, cities in: Fairmont, 171; Faribault, 171; Lake Lillian, 171–72; Mankato, 72; Minneapolis, 147–48, 153; Truman, 171; Winnebago, 171
Minnesota Bible College, 148
Minnesota Geological Survey, 171–72
Minnock, Thomas J., 19
Missouri, cities in: Joplin, 140; Saint Louis, 140, 177
Mitterrand, François, 235
Mobil Oil, 99
Moineau, Henri, 184, 189–89
Monk, Noel J., 96–97
Monk Oil Company, 97
Monongahela Valley Republican, 13
Montana, cities in: Butte, 166; Townsend, 91
Montana Oil Journal, 154
Moosman, John, 104–5
Morris, Joe, 140

Morrow County, Ohio, 96–97
Mosher (dowser), 97
Mount Davidson Oil Company, 18
Mu (lost continent), 27
Muensterberg, Hugo, 73
Mumbai, India, 118
Munger, Ellwood, 173
Murphy (spirit medium), 18
Mussolini, Benito, 113
Mussolini, Rachele, 113
Mystic Oil, 38

Nagurski, Bronko, 205
Nash, Lloyd N., 174
National Doodlebuggers Convention, 206
National Oil Journal, 144
Natural Gas, 136
Nature, 139
Nebraska, cities in: Beatrice, 186; Campbell, 160–61; Lincoln, 159; Omaha, 160–61, 183; Red Cloud, 186
Nelson, Henning, 171–72
Nelson, Walter J., 102–3
Nevada, cities in: Las Vegas, 54; Reno, 160, 162; Virginia City, 18
New and Unusual Methods of Petroleum Exploration, 163
New Jersey, cities in: Atlantic City, 125; Jackson's Mill, 169; Lakewood, 168; Nutley, 116; Passaic, 116
New Mexico, cities in: Albuquerque, 39; Aztec, 216, 226; Belen, 39; Dayton, 79; Deming, 225; Grants, 98; Hobbs, 96; Roswell, 79, 216; Silver City, 224–25
New York, cities in: Albany, 92; Barden Brook, 18; Belmont, 18; Bolivar, 18; Clifton Springs, 13; Elmira, 10; Fredonia, 17; New York, 60, 116–17, 136, 192, 207–9, 218; Port Jervis, 116; Rochester, 12; Syracuse, 18
New York Bureau of Securities, 208
New York Journal, 208

New York Times, 218
Newsweek, 9
Newton, Nan O'Reilly, 208–9
Newton, Sharon Chillison, 215, 223, 226
Newton, Silas Mason, 207–26
Newton Oil Company, 218
Nobel, Ludwig and Robert, 81
North Dakota, cities in: Edgeley, 91; Ellendale, 163; Nesson, 89; New England, 89; Ray, 89; Robinson, 88; Tioga, 89; Turtle Lake, 165–66; Valley City, 89
Nutt, Roy, 96–97

Occidental Oil Company, 99
Ogle County, Illinois, 177
Ohio, cities in: Cleveland, 21; Columbus, 96; Columbus Grove, 71; Delphos, 71; Findlay, 70; Lima, 70; Marysville, 78; Miamisburg, 70; Portsmouth, 205; Taylorsville, 70; Van Wert, 70–71
Oil & Gas Journal, 84, 87, 124, 134, 179, 188, 209
Oil Compass, 186
Oil Reporter, 212
oil smeller, 68–70
Oil Springs, Ontario, Canada, 109
Oil Thermometer, 186
oil witch, 68, 179
Oilometer, 151
Oklahoma, cities in: Altus, 132–34; Bartlesville, 141; Chickasha, 164; Edmond, 200; Elk City, 134; Kingfisher, 202; Oklahoma City, 133–34, 164, 200–201; Shawnee, 206; Tulsa, 134, 188, 202, 207; Weatherford, 206
Oklahoma, counties in: Kingfisher, 202; Okmulgee, 41, 79; Pontotoc, 41; Pottawatomie, 205; Washington, 30
Oklahoma, oil fields in: Cushing, 83; Healdton, 133; Muskogee, 80; West Edmond, 200–201, 205–7
Oklahoma History Center, 200–201

Olson, Carl, 148–51
Olson, David Eugene, 147–55, 157, 178
Ontario Department of Mines, 109
Ordóñez, Ezequiel, 81
Oregon, cities in: Corvallis, 155; Cottage Grove, 147–53, 155; Eugene, 148–52, 155, 178; La Grande, 154; Monmouth, 125; Oregon City, 126; Portland, 125, 155; Pratum, 126; Salem, 149
Oregon, counties in: Lane, 149; Marion, 126
Oregon State University, 151
Oriental Refining Company, 209, 219
Oriskany Sandstone, 9
Orlando Sentinel, 60
Oscilloclast, 138
Osty, Eugene, 73
Oswald, Lee, 55
Owens, Jimmy, 47
Owens, William, 27

Pairol (dowser), 120
Parapsychology Foundation of San Diego, 54
Partridge, William, 118
Paterson, William Innes, 185–87
Péan, Pierre, 230, 284n74
Pearson's Magazine, 138
Pennsylvania, cities in: Bradford, 159; Coudersport, 75; Hydetown, 7; Oil City, 5; Petroleum Center/Centre, 7, 67, 73; Philadelphia, 156; Pithole, 67; Pittsburgh, 158; Pleasantville, 15, 17; President, 16–17; Tidioute, 67; Titusville, 5, 11–14, 17, 67, 73, 75; Watrous, 10
Pennsylvania, counties in: Armstrong, 12; Clarion, 16; Venango, 67
Pennsylvania, oil fields in: Clarion, 16; Gaines, 10; Millerstown, 12; Parker, 12; Pithole, 12, 67; Pleasantville, 12, 15–16
Pennsylvania Geological Survey, 8, 19
Penrose, Evelyn, 118–20

perpetual motion machine, 163–64
Perron Oil, 168
Perry, Clifford, 51–52
Perry, George W., 166–69
Perth, Australia, 119
Pesenti, Carlo, 228–29
Petroleum Age, 200
Petroleum Maatschappij Salt Creek, 81
Petroleum Magnet, 145
Petroleum Times, 115, 118
Petroleum World, 117
Petrolia oil field, 80
Petrolometer, 151, 155, 157–58
Pflock, Karl, 226
Philadelphia Museum, 92
Phillips, Glen, 185
Phillips Petroleum, 59–61, 212
Pierre Shale, 76–77
Pinay, Antoine, 229, 231
Pittsburgh Post, 74
Pittsburgh-Louisiana Oil Company, 158
Placid Oil, 49
Plummer, A. Newton, 208
Pocatello, Idaho, 90
Pogson, Charles A., 118
Poland, cities in: Poznan, 72; Pudewitz, 72
Pope, J. E., 145
Porro, Cesare, 81
Portland Oregonian, 147
Post, Wiley, 180
Postal Service, 58, 147, 165
Pratt, Wallace, 84
Pravda, 121
Producers Associates Inc., 174
Producers' Oil Company, 130–31
Producer's Oilwell Supplies Ltd., 188–89
Producing Oil Fields Ltd., 188
Prokop, Otto, 116
psionics. *See* radionics
psychic oil exploration, 21–61
psychometry, 246n13
Purchalia, Stanislas, 116
Purdy, Ken, 223
Putnam, I. M., 164

Pyle, Denver, 57–58
Pyle, Earl, 61, 100–102
Pyle, John, 101
Pyle, Sarah, 61, 101
Pyle, Willie, 61, 101

radio perceptionist, 119
Radio-Electro-Magnetic Teledetection, 184
Radiograph, 165, 187–88
Radiologometer, 173
radionics, 4, 138
Radioscope, 145, 164–65
Radioscope Laboratories, 164
Raft, George, 57
Ramona Oil Company, 180
Rand, Sally, 209
Rapid City, South Dakota, 161
Rayburn, Don, 120
Raynor, Morris, 31
Reader's Digest, 9
Red Feather Oil and Gas, 61
Reed, William, 125
Reese, Bert (Berthold Riess), 72–74
Reflexophone, 138
Regis, 184, 189
Register (Eugene, Oregon), 149, 153
remote viewing, 61
Renovo Daily Record, 8
Richfield Oil, 99
Rifle machine, 138
Ringle, Cecil, 30, 38–39
Rio Arriba County, New Mexico, 39
Roberts, Kenneth, 64, 92–95
Robinson, James, 18
Rockefeller, John D., 21, 72–74
Rocky Mountain Petroleum Yearbook, 212
Rogers, Will, 180
Roosevelt, Franklin, 56
Rossville, Georgia, 56, 58
Ruby, Jack, 55
Runyon, Damon, 215
Rush, Joe, 31
Russian Academy of Sciences, 121
Rynd, John, 12

Sage, W. A., 148–49, 152
Saint Johns Review (Portland, OR), 126
Saint Louis Republic, 22
Salem (Oregon) *Capital Journal*, 149
Sam Davis Oil Company, 28, 30
San Andres Formation, 79
San Francisco Chronicle, 218
San Saba Exploration, LLC, 42
Sanderson, Eugene C., 149, 152
Saturday Evening Post, 102–3
Saturday Review, 218
Schermuly, Philipp, 190–93
Schermuly Polarizer, 190–93
Schiaparelli, Giovanni, 241–42
Schmid, Adolf, 194, 197–98
Schmidt, N. G., 121
Schröder-Stranz, 194
Schwartz, H. H., 145–47
Schwartz, Stephan, 95–96
Science in Russia, 121
Scientific American, 136, 139, 192
Scott, Samuel, 185
Scully, Alice, 220
Scully, Frank, 209, 215–20, 223
Scully, Moreen, 215
Sears, L. E., 136
Securities and Exchange Commission (SEC), 58, 162–63, 166, 172–74
seismic reflection, 157, 199–200, 241
seismic refraction, 199–200, 241
Selma, Alabama, 28
Seventh Sense, The, 92
Sharpe (psychologist), 21
Sharpe, William, 159
Shaw, John R., 55, 57–59
Shell Oil, 81, 114
Sherman, Harold, 59, 224
Sherwin, Charles, 145–47
Shoaf, John, 86
Silliman, (Prof.) Benjamin, 65
Sinclair, Harry, 5
Sinclair, Upton, 138
Sjögren, Hjalmar, 81
Skłodowska-Curie, Maria, 184
Slick, Dewey, 102

Slick, Tom, 158, 187
Smith, Billy, 12
Smith, P., 67
Sochevanov, N. N., 121
Solovkin, A. P., 121
South Australia, cites in: American Beach, 188; Mount Gambier, 188
Southwest Minnesota Oil Exploration Company, 172
Southwest Research Institute, 158
Southwestern Oil Company, 79
sphygmobiometer, 138
Spindletop well, 23, 77, 125, 129, 143
Spirit Oil Company, 18
spiritualism, 11–19; decline of, 19
Spitz, Samuel, 151, 155–58
Spitz flight recorder, 156
spitzascope, 155–56
Spraggett, Allen, 139
Stalin, Joseph, 120
Standard Oil Company, 72–74, 82, 170, 190
Stanford Research Institute, 61
Steinbuchel, Max, 187–89
stigmata, 58
Stony Mountain Oil and Gas Company, 183
Stork, Clarence, 184–85
Stygian rays, 155
Sugrue, Thomas, 42
Sun Oil, 244
superstition, 5–6, 47, 78
Swann, Ingo, 60–61
Swift, Tom, 137
Switzerland, cities in: Altishofen, 112; Geneva, 112; Saint-Prex, 112; Semsales, 112; Tuggen, 112, 193
Switzerland: Vaud Canton, 112
Swoboda, Alois P., 89–91
Sylvania Electric Company, 92
synchronal compass, 178–79

Tampico oil fields, 36
Tarbell, Ida, 82
Tarot cards, 63

Tedeschini, Marco, 227
Telegram Oil Company, 34
Telford, George H., 183
Telford, J. Lyle, 181–83
Teller, Henry, 21
Teller, John, 21
Teller, Stella, 21–23
Terrestrial Wave Detector, 151, 178
Texaco (the Texas Company), 83, 131, 209
Texas, cities in: Austin, 87; Beaumont, 23–24, 77, 125, 127, 143; Breckenridge, 144; Burkburnett, 144–45; Cleburne, 28; Colorado City, 87; Comyn, 28, 30, 249n35; Corpus Christi, 142; Corsicana, 25–26; Dallas, 42, 45, 54–55, 200; Denton, 146; Desdemona, 28–31; Electra, 79, 144; Fort Worth, 87, 145–46, 167; Houston, 23, 26, 127–29, 132, 144; Humble, 127; Kilgore, 44; Laredo, 87, 141; Lockhart, 85–87; Longview, 48; Luling, 31–32; Marfa, 100; McAllen, 206; Mexia, 26, 142–45, 159; Midland, 55; Ranger, 30, 144; Rockwall, 134; San Angelo, 87; San Antonio, 86–87, 145, 158, 164; San Saba, 32, 42, 249n35; Sanderson, 23, 100; Smiley, 32; Springfield, 26; Uvalde, 23; Wichita Falls, 23, 86
Texas, counties in: Bosque, 34–35; Caldwell, 85, 87; Coleman, 41, 93; Comanche, 39, 41; Crockett, 35, 41; Culberson, 100; Gonzales, 31–32; Limestone, 131; Matagorda, 128; Mills, 55; Nolan, 54; San Saba, 27, 32–33, 37–38, 42–43, 93; Smith, 40; Wichita, 38–39; Zapata, 141
Texas, oil fields in: Batson, 129; Big Hill, 128; Burkburnett, 133; Corsicana, 26; East Texas, 40, 46–47; Electra, 75, 79–80; Humble, 86, 127, 129–32; Ireland-Conser, 54; Laredo, 141; Luling, 31–32; Mexia, 131; Ranger, 80; Sour Lake, 129–30; Spindletop, 24; West Columbia, 129
Texas sharpshooter fallacy, 95, 261n46
There Is a River, 42
This Gay Knight, 218
Thompson, Andrew, 74
Thrash, Day Matt, 28
Tikal, Guatemala, 92
Time, 203, 218
Tinsley oil field, 48
Tippit, 79
Titusville Herald, 14–15, 67, 87
Tom Swift and His Great Oil Gusher, 137
Tom Swift and His Motor-Cycle, 137
Topeka Daily State Journal, 179
Toronto National Post, 184
Townley, Arthur C., 86–90
Trenton Limestone, 51
Tribune (Bismarck, North Dakota), 89, 147
Triumph Oil and Gas, 183–84
Triumph Oil Company, 77
Tropel, Jean, 229, 231, 236
True, 220, 223
Trumbull, M. C., 144, 158–59
Truman, Harry, 59

Union des Pétroles d'Oklahoma, 81
Union Oil Company, 48, 82
United Oil Company, 77
United Producing Company, 97
United States Geological Survey, 100, 242
United States Marine Corps, 64
University of California, 156
University of Chicago, 156
University of Denver, 217
University of Halle, 114
University of Heidelberg, 138
University of Kansas, 59
University of Lausanne, 111
University of Life Church, 53
University of New York, 54
University of Oregon, 149–50
University of Panama, 54
University of Paris, 111
University of Texas, 98, 141, 176

University of Vienna, 54
University of Wisconsin, 169
Upshur County, West Virginia, 176
Urquiza (doodlebugger), 120
Utah, cities in: Murray, 56; Ogden, 90, 157; Salt Lake City, 90
Ute Indians, 75
Uvarov, L., 121

Vandergrift, 73
Variety, 215–16
Viet Cong, 64
Vincent, Mordelo, 120
Violet, Jean, 228–29, 231, 233
Virginia Beach, Virginia, 41
Virginia City Territorial Enterprise, 18
Vitapathy, 17
Von Tüköry, Charlotte, 110, 115
Von Uslar, Rafael Perfecto, 115–16

W. W. Keeler and Sons, 61
Wadley, J. K., 98–99
Walker, Clint, 57
Walther, Johannes, 114
Washburn, Don, 53
Washington, DC, 48
Washington (state), cities in: Seattle, 166; Spokane, 146–47; Tacoma, 185; Yakima, 157
Washington Post, 218
Washington State College, 54
Wasserman (oil operator), 73
water smeller, 69
Water Unlimited, 92, 94
Watson, D. M., 125–26
Watson, Elizabeth Lowe, 12–14
Watson, Hugh, 183–84
Watson, Jonathan, 11–14, 17, 19
Wayne County, New York, 159
Weber Sandstone, 209
West, Annie Webb (Annie Buchanan, Annie Jackson), 25–27, 144
West, Clark, 25–27
West, Mae, 54
West Virginia, cities in: Belmont, 104; Cairo, 104; Parkersburg, 104–5; Saint Marys, 104
Western Union, 34
White County, Arkansas, 98
Wichita Beacon, 132
Wichita Daily Eagle, 87
Wiesinger (oil operator), 73
Wigglesworth Pathoclast, 138
Wilcox Sand, 204
Wilhelm (Kaiser), 115
Williams, Robert, 94
Williamson, Harold F., 81
Wilmott, Curtis, 37–38, 42
Winchell, Walter, 209
Windsor Magazine, 195
Winter (spirit medium), 18
Wireless Spitzascope Company, 155
Wisconsin, cities in: Antigo, 170; Menomonie, 70; Rhinelander, 169
Wisconsin Oil and Development Company, 169
Wisconsin Petroleum and Development Company, 169
Wisconsin Railroad Commission, 169–70
Wittich, W. F., 144
Woodward, C. H., 170
Woolsey, Vern, 31–32
Wright, William ("Dan DeQuille"), 18, 68
Wucherer, G. F., 180
Wyoming, oil fields in: Dutton Creek, 214–15; Salt Creek, 81
Wyrick, Madison, 34–35

Yale College, 65
Yale Lock Company, 50–51
Yavapai Oil Company, 224
Young, James M., 200–201, 204–5
Yuma County, Arizona, 36

Zachary, William Henry, 85–87, 89
Zhytomyr, Ukraine, 202
Zistersdorf oil field, 116–17
Zuck, Rochelle Raineri, 246–47n13

DAN PLAZAK is a retired geologist and engineer, a graduate—he prefers the term survivor—of Michigan Tech and the Colorado School of Mines. He is the author of a history of swindling in the mining industry, *A Hole in the Ground with a Liar at the Top*. His principal interest is in oddball topics neglected by other historians. He lives in Denver, Colorado.

(Author photo by Nicole Sproveri)

Printed in the USA
CPSIA information can be obtained
at www.ICGtesting.com
LVHW090543131023
760935LV00003B/321